DNA FOR ARCHAEOLOGISTS

We dedicate this book to Tessa and Elsie,
our mitochondrial legacies.

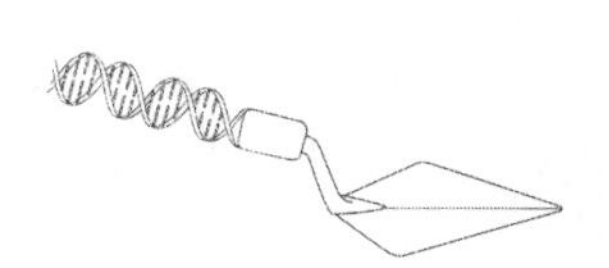

DNA FOR ARCHAEOLOGISTS

Elizabeth Matisoo-Smith and K. Ann Horsburgh

Walnut Creek, California

LEFT COAST PRESS, INC.
1630 North Main Street, #400
Walnut Creek, CA 94596
http://www.LCoastPress.com

ISBN 978-1-59874-680-8 hardback
ISBN 978-1-59874-681-5 paperback
ISBN 978-1-59874-682-2 institutional eBook
ISBN 978-1-61132-482-2 consumer eBook

Library of Congress Cataloging-in-Publication Data:

Matisoo-Smith, Elizabeth.
DNA for archaeologists / Elizabeth Matisoo-Smith and K. Ann Horsburgh.
p. cm.
Includes bibliographical references and index.
ISBN 978-1-59874-680-8 (hbk. : alk. paper) — ISBN 978-1-59874-681-5 (pbk. : alk. paper) — ISBN 978-1-59874-682-2 (institutional eBook) — ISBN 978-1-61132-482-2 (consumer eBook)
1. Human remains (Archaeology)—Analysis. 2. Human molecular genetics. 3. Mitochondrial DNA. 4. Biomolecular archaeology. 5. Archaeological chemistry. 6. Archaeology—Methodology. I. Horsburgh, K. Ann, 1975- II. Title.
CC79.5.H85M38 2012
930.1—dc23

2012022651

Printed in the United States of America

∞™ The paper used in this publication meets the minimum requirements of American National Standard for Information Sciences—Permanence of Paper for Printed Library Materials, ANSI/NISO Z39.48–1992.

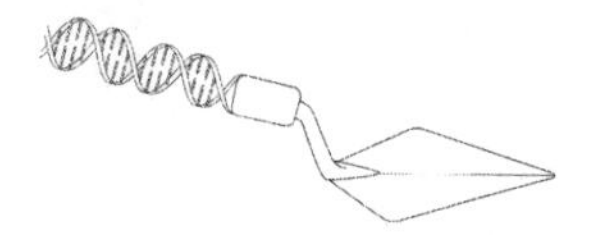

Contents

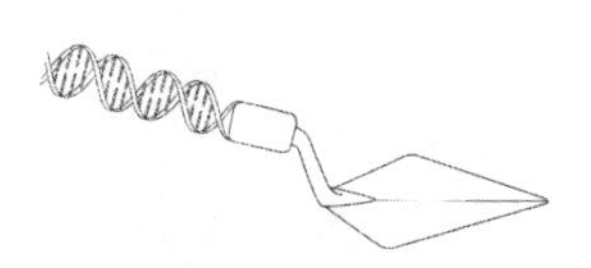

Illustrations

Boxes

Figures

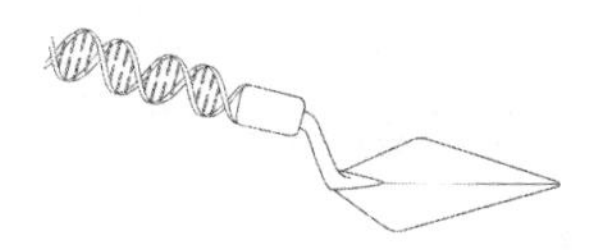

Acknowledgements

We would particularly like to thank and acknowledge our mentors, Roger Green and Richard Klein, both archaeologists who encouraged us to think big and think holistically. We hope we have not let you down. Additional recognition and thanks go to Dave Lambert who took an anthropologist, untrained in molecular methods, and introduced her to the molecular biology laboratory, the remarkable adrenaline rush of last-minute grant writing, and the encouragement to "just do it."

Over the years, we have been privileged to spend much time in the field with our archaeological colleagues David Addison, Graham and Margaret Avery, Tom Huffman, Herman Mandui, Mark McCoy, Jose Miguel Ramirez, Thembi Russell, Andrea Seelenfreund, Glenn Summerhayes, and Richard Walter. We thank them for their company, patience, support, and the numerous discussions we have had about the various ways to reconstruct the past. We also wish to thank the many archaeologists with whom we have collaborated and who have provided us with samples—we do appreciate the hard work that goes into obtaining them! Thanks to our colleagues who have worked with us in the lab on various aspects of aDNA and molecular anthropology, particularly Craig Millar, Judith Robins, Michael Knapp, and Jo-Ann Stanton.

Thanks to Vivian Ward for her wonderful illustrations, and to Craig Millar for all of his assistance and ideas regarding illustrations. We would also like to acknowledge and thank Anna Gosling and Julie Matisoo for their excellent editorial skills and assistance.

Many helpful comments were received from two anonymous reviewers, and we particularly thank George Armelagos who reviewed the manuscript and made significant comments and valuable suggestions. We are substantially indebted to Mark McCoy who enthusiastically played the role of a "typical archaeologist" as he read multiple drafts and pointed out when we assumed too much or too little of our target audience.

We are forever thankful for Brent and Mark who put up with us and without us during the writing process and who have been endlessly supportive of our work.

CHAPTER ONE

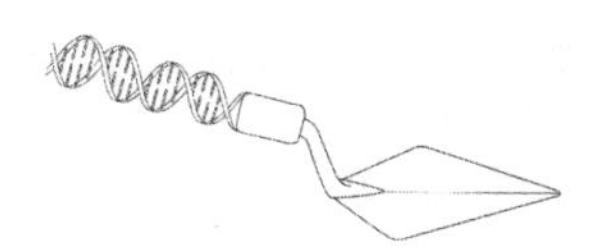

Why Read This Book?

INTRODUCTION

DNA studies are beginning to become part of everyday life. Once hidden away in the realms of the science laboratory, personal DNA information is now available to anyone with access to the internet and a credit card. Just Googling "DNA ancestry" results in over 8.5 million hits. By providing a DNA sample to any number of DNA analysis companies, you can now not only identify your ancestors for approximately US$500, you can "take charge of your health," and have your DNA analyzed to learn if you are carrying inherited disease markers. In addition, just in case there was any doubt, they could also tell you what color your eyes are likely to be, how curly your hair is, or if you have wet ear wax! Generally, we could consider this a good thing. If DNA is accessible, then it is no longer scary, people might read a bit more about it, and they would begin to understand what can and can not be done with DNA technology. This information might help consumers negotiate issues that they face in their daily activities. For example, DNA topics related to everyday life include among others: 1) What is genetic modification, and how will it affect my food?; 2) How do stem cells work, and should I be storing my child's cord blood for the future?; or 3) What does it really mean if I am carrying a genetic marker that is associated with breast cancer? Similarly, if people understand how closely related all human beings are to each other, then we might be able to dispel the negative and destructive concepts of racial purity and/or superiority.

Unfortunately, not all who engage superficially with DNA technology and information will fully understand or correctly interpret their results.

They may not consider the ramifications of finding out that they are carrying markers known to be associated with particular diseases, or they may not appreciate that disease development is complex and that having an increased likelihood of contracting a disease does not necessarily mean that they will. This is the drawback of the mass appeal, public access, and engagement with DNA.

Just as genetic data can be misleading to a member of the general public, who has found out through his Y chromosome analysis that he is a direct descendent of Charlemagne (like 99% of European males, Rohde et al. 2004), they can also be misused as they were for political purposes during the eugenics movement of the early twentieth century. In the same way, genetic information can be equally misused and misinterpreted in archaeology. Since the mid-1980s, DNA has had a significant impact on archaeological theory and practice. Studies of modern human **mitochondrial DNA** (mtDNA) and Y chromosome variation (among other markers) have been used to theorize about population migrations, structure, expansions and replacements, and kinship and social organization.

The discovery that DNA still remained in, and could be obtained from, archaeological remains brought it directly into the archaeologist's tool kit. Not only can archaeologists obtain direct information about the biological affinities of the individuals recovered from a site, they can determine their sexes, study the animal remains associated with them, identify possible diseases, and address issues about family relationships or population replacement within a site. However, genetic analyses, generally, and ancient DNA (aDNA) more specifically, are relatively new techniques, and as such, there have been a number of teething pains in their application. Early researchers did not consider or were unaware that there are major problems resulting from contamination of archaeological samples with modern DNA. It has been argued that aDNA has "come of age" (Nicholls 2005) partly because we now have standard protocols and requirements for aDNA laboratories and research. Despite this fact, papers that do not meet these standard requirements are still being published, and unfortunately, they are being published in archaeological and anthropological journals. This is perhaps happening due to the constantly changing technology, and as research has become increasingly specialized, it is difficult enough for molecular biologists to keep up with new techniques, applications, and debates—let alone for archaeologists who are focused on their own literature.

Traditionally, anthropology has been split into subfields. In the American system, these include archaeology, biological or physical anthropology, social or cultural anthropology, and anthropological linguistics. The degree to which these and other subdisciplines interact varies by university and even national anthropological tradition. These main subdivisions have, however, increasingly been further broken down into specialist subfields within fields. They may be based on geographic regions, time frames, or methodological emphasis such that archaeologists, for example, often self-identify as specific regional archaeologists, historical archaeologists, or landscape archaeologists. Within biological or physical anthropology, a similar process has taken place with researchers focusing on paleoanthropology, primatology, or similar specific subfields. Trained in the four-field, "holistic" approach to anthropology, the authors of this volume value the tradition but also recognize that the reality and politics of many departments make such an approach difficult. Yet, given that the title of this book is *DNA for Archaeologists*, we assume that most readers will recognize that biological methods can be useful in addressing archaeological questions.

Because this book is for archaeologists, we do not focus in detail on the specific laboratory methods or intricacies of DNA research. Instead, we hope to introduce archaeologists to the basics of DNA, so that they can understand both the powers and pitfalls of genetic data and the application of such data to issues of interest to them. We also hope to provide enough specific information so that archaeologists can use or even contribute to DNA research from a position of confidence and make informed decisions about molecular research that addresses issues of prehistory and history.

We are molecular anthropologists. The term *molecular anthropology* is often used synonymously with anthropological genetics; however, we consider the emphasis both significant and important. In molecular anthropology, the stress is on the anthropological dimensions of the research, which is distinct from work that is theoretically and practically situated in the field of genetics. As anthropologists, we are interested in understanding and explaining human diversity, be it biological, linguistic, or cultural. Clearly, part of the explanation for observed diversity is related to human history. Studying the material remains of the human past is the realm of archaeological research, but molecular anthropologists can study human history through DNA—what we inherit from our parents and other ancestors. We also know, however, that DNA is not the blueprint for an organism because it is clearly the interaction between genetics/DNA and the environment that determines the characteristics of the organism. For humans, this environment can be biologically as well as culturally determined. Thus, we recognize this complexity and apply a biocultural perspective to the understanding and explanation of human diversity. To reiterate, an anthropological perspective is important when looking at and explaining human variation. Variables such as whom we marry and where we live will have an impact on the patterns of diversity that we see. Chance factors, such as genetic drift, which might be related to population migrations or to environmental catastrophes, may also influence those patterns. Therefore, if the researcher does not appreciate or understand the possible historic scenarios or consider the interactions between biology and culture, the patterns of genetic variation reflecting human behavior through time and across space may be misinterpreted. Regardless of our views about the importance of grounding the interpretation of genetic data in an anthropological framework, archaeologists are increasingly faced with genetic data, and for that reason, they need to know the basics about DNA and its properties.

A Brief History of Molecular Anthropology and Ancient DNA

Unlike the field of archaeology, which has a long past, one must remember that it was just over one hundred years ago that Gregor Mendel's research on pea plants and the mechanisms of inheritance were rediscovered, marking the birth of the field of genetics (Correns 1900; De Vries 1900; Mendel 1865). The structure of DNA, the hereditary material, was only discovered in 1952 (Watson and Crick 1953a and 1953b). The term molecular anthropology was first used in 1962 when Emile Zuckerkandl spoke at a conference organized by Sherwood Washburn about research that was being undertaken by biochemists who were applying their tools to anthropological questions, specifically comparing human and gorilla hemoglobin (Zuckerkandl 1963). It can be argued, however, that true molecular anthropology, situated within anthropology and practiced by anthropologists, begins with the work of Vincent Sarich and Allan Wilson (1967a, 1967b; Wilson and Sarich 1969). Sarich was undertaking his PhD research in the Department of Anthropology at the University of California, Berkeley, under the supervision of Wilson, who had recently been appointed to a position in the Department of Biochemistry, and Washburn, who had been at the university for some time. It was Sarich and Wilson's work that applied and further developed Zuckerkandl and Pauling's (1965) theory of a "molecular clock" to genetic research and thus brought the time component of genetic change to anthropological debates. These were the first steps toward bringing the worlds of genetics and archaeology together.

Allan Wilson carried on his pioneering research in molecular evolution and molecular anthropology, both theoretically and through technical developments, until his untimely death in 1991 (Cann 1993). He trained and mentored a generation of researchers who ultimately established molecular anthropology as a major field of research and went on to develop their own labs around the world. In the mid-1980s, geneticists at Allan Wilson's lab undertook the first real aDNA study when they extracted and sequenced DNA from an extinct animal. The quagga (*Equus quagga quagga*) was a zebra-like species that became extinct in 1883 largely as a result of big game hunting. To collect DNA, they removed muscle and connective tissues from the skin of a stuffed quagga stored at the Museum of Natural History in Mainz, Germany, and sequenced 229 base pairs of quagga mtDNA (Higuchi et al. 1984). They concluded that the quagga was indeed most closely related to the plains zebra and not the domestic horse, as some had thought. They further noted, in what may have been one of the most famous understatements of science, "[i]f the long-term survival of DNA proves to be a general phenomenon, several fields including paleontology, evolutionary biology, archaeology and forensic science may benefit" (Higuchi et al. 1984:284).

In 1985, Svante Pääbo, who would later become another of Wilson's doctoral students, investigated the possibility of obtaining DNA from twenty-three Egyptian mummies from a range of European museums. Working again with skin and other connective tissue samples, he was able to obtain DNA from only one, an infant boy dated to 2400 BP (Pääbo 1985). The collection of the mummy sample was undertaken in what we know today to be unacceptable conditions—Pääbo was not even wearing gloves (Gitschier 2008)—and the results were not replicable (Pääbo et al. 2004). Therefore, Pääbo was probably the first person, but by no means the last aDNA researcher, to sequence contamination; in this case, it was probably his own DNA. Although in doing so, he helped advance the field nevertheless. He later went on to establish one of the premier aDNA labs in the world at the Max Planck Institute for Evolutionary Anthropology in Leipzig, Germany.

While several members of Wilson's lab were working on ancient tissue samples, others were investigating a possible new tool for understanding population histories. That new tool was mtDNA. Wilson recognized that one of the difficulties of studying nuclear DNA markers to reconstruct historical relationships was that, due to recombination, the DNA gets shuffled in each generation (see chapter two). Mitochondria, on the other hand, are passed down only through the maternal line, and they contain their own DNA. Previous research indicated that mtDNA accumulates mutations rapidly and, therefore, is particularly useful for studying variation within a species (Brown et al. 1979). If researchers could extract mtDNA from tissues, then they should be able to track the history of female descent through time and space.

Doctoral students at the Wilson lab conducted studies on mtDNA from a range of animals, including fish, birds, and nonhuman primates (Kocher et al. 1989; Meyer and Wilson 1990; Meyer et al. 1988; Meyer et al. 1989; Meyer et al. 1990). They published their results in top academic journals; although there was little or no apparent awareness of or interest in their work outside the field of molecular biology. It was the two students working on human mtDNA who produced results that attracted the popular press. In 1987, Rebecca Cann, Mark Stoneking, and Allan Wilson published a paper describing mtDNA variation in a large human sample representing populations from around the globe. Their results indicated that all human mtDNA could be traced back to a single common ancestor who lived approximately 200,000 years ago in Africa (Cann et al. 1987). This finding became known as the "mitochondrial Eve hypothesis," and it made the cover of *Time Magazine*.

At about this same time, Kerry Mullis, another Berkeley-based biologist, was thinking about how DNA replicated itself in a cell when he had an epiphany. He realized that it could be possible to synthesize DNA artificially (Mullis and Faloona 1987; Mullis et al. 1986). By identifying and targeting a particular region of interest, one could harness the process of DNA replication and create thousands of copies of that fragment of DNA reasonably quickly in the lab. This process, known as the Polymerase Chain Reaction, or PCR, revolutionized molecular biology.

Prior to the development of PCR, researchers had to use fairly complicated methods to extract nuclear or mtDNA using microorganisms to replicate it (the process known as cloning) and create enough DNA to analyze for sequence variation. This process required significant laboratory equipment to complete, and it was slow and expensive. Using PCR, researchers could quickly and selectively copy a particular DNA fragment from within the entire genome, providing enough to be directly analyzed (for example, for the actual sequence of DNA nucleotides, A, T, C, and G, to be read) without cloning.

Polymerase Chain Reaction—which copies DNA by combining the sample with the appropriate mix of chemicals and buffers and taking it through a series of temperature fluctuations—was initially performed using a series of water baths. Soon, however, PCR machines, which are basically computer-controlled heating blocks, were manufactured and became relatively affordable. With PCR, almost anyone could amplify DNA with minimal laboratory investment or experience. As a result, molecular biology became accessible and could be done just about anywhere—even in a hotel room (Dizon et al. 2000).

The ability to extract DNA from bone and, therefore, archaeological remains was the final step in bringing together molecular biology and archaeology. The first study to describe the extraction and analysis of aDNA from archaeological bone samples was conducted by Erika Hagelberg and her colleagues at Oxford (Hagelberg et al. 1989). They amplified mtDNA from a series of human burials recovered from English cemeteries dating from 4810 BP to the seventeenth century and a 5,450-year-old femur from a cave burial in the Judean Desert. These early studies produced great excitement as researchers realized that significant archaeological questions could be addressed through analysis of DNA obtained from dateable archaeological samples. It would be several years however, before researchers realized that the contamination of ancient samples with modern DNA was a significant problem for aDNA studies. While we now acknowledge the inevitability of contamination, its identification and the authentication of true, or endogenous, aDNA continues to be a major issue that aDNA researchers must contend with.

Thus, by the end of the 1980s, molecular research into human history and variation was established and making major contributions to anthropological research. Technological developments, including the development of an automated DNA sequencing machine in 1987, had significantly reduced the cost of obtaining DNA data, and a world of possibilities opened up. It was at this point that the scale of genetic research was about to take off, as in 1990 the Human Genome Project (HGP) was announced. The main goal of the HGP was, over a period of thirteen years, to determine the sequence of all three billion DNA bases of the human genome and to identify the twenty thousand or more genes in order to better understand the genetics of human disease (Human Genome Project 2012). This announcement was soon followed by a proposal for the development of a subgroup of the HGP, the Human Genome Diversity Project (HGDP). While the goal for the HGP was to create a single amalgam sequence of human DNA, a kind of mosaic of the DNA of several individuals, the HGDP was proposed to address the issue of genetic diversity in human populations. The leaders of the HGDP made a plea to scientists around the world to "collaborate now in collecting sufficient material to record human ethnic and geographic diversity before this possibility is irretrievably lost" (Cavalli-Sforza et al. 1991:490).

The HGP was completed in 2003, two years ahead of schedule; though draft sequences were released in 2001 by the HGP and by a competitor group led by Craig Venter and his company, Celera Genomics (McPherson et al. 2001; Venter et al. 2001). Unfortunately, the HGDP was not so successful. Over the first few years of its launch, the HGDP hosted four symposia focused on sampling, anthropological, organizational, and ethical issues. It was the last of those issues that proved the major stumbling block for the project. Although the ethical issues raised by the HGDP are discussed at length in chapter four, in short, indigenous communities and others raised numerous concerns about issues, such as lack of consultation, bio-piracy, and scientific racism (Hasian and Plec 2002). The project then ground to a halt trying to deal with the criticisms, and in 1994, the funding agencies, the United States National

Institutes of Health, the National Science Foundation (NSF), and the Department of Energy, all requested that the Research Council of the United States National Academy of Sciences convene a special committee to assess the feasibility and ethical implications of the HGDP. This committee deliberated for several years. Finally in 1997, it made a public recommendation that the project could proceed, but only if it adhered to a strict code of ethical guidelines.

Funding for the proposed HGDP collection of cell lines from "anthropologically significant" populations around the world was never secured, but a number of researchers who already had cell line samples contributed some of those to a central collection now housed at the Centre for the Study of Human Polymorphisms (CEPH) in Paris. As of 2005, this HGDP collection consisted of 1,064 cell lines from fifty-two populations from around the world. DNA samples from this collection are made available, free of charge, to any researcher who is undertaking research that is nonprofit as long as they agree to send their data back to the CEPH database where they can be made publically available (Cavalli-Sforza 2005).

There are major limitations, however, to the samples that are currently in the HGDP collection. Most importantly, the total number of samples is small, and it does not provide a good coverage of populations around the world. In particular, samples from India and Polynesia are completely missing, and even Europe, northern Asia, the Americas, and Oceania are poorly represented (Cavalli-Sforza 2005). While the HGDP may never be able to undertake targeted field sampling to fill in these gaps due to concerns about the nature of the research and the type of samples being collected (see discussion in chapter four), another major collaborative project is currently underway collecting data specifically to help address questions about human origins and population dispersals across the globe.

In 2005, National Geographic launched the Genographic Project. The project was designed to be a five-year, international research effort to collect and analyze DNA samples from around the world. Unlike the HGDP, it specifically focuses on anthropological issues and tracking human population origins and migrations by studying only mtDNA and Y chromosome diversity. It is nonmedical, nonprofit, and nongovernmental; and it is hoped that the project personnel have learned from the mistakes of the past when it comes to the ethics of working with human subjects who may have a very different way of viewing genetic research and technology (Wells and Schurr 2009). The project has taken a two-pronged approach to data collection. On one side, the various investigators involved in the Genographic Project undertake fieldwork and collaborate with indigenous communities to collect DNA samples.

The aim is the collection of up to 100,000 new, anthropologically significant, samples making it the largest survey of its kind. The second approach to data collection is through public participation. At the time of publication, individuals could log onto the Genographic Project's web site provide credit card details, order a kit to collect their own DNA (through a cheek swab), send it in for analysis, and get their results. They could then choose to include their results in the Genographic database of DNA data. Once all of the Genographic data are analyzed, the projects aim is to make the data available to other researchers and to the public through the Genographic database (though anonymity of participants is maintained).

The issue of reproducibility of and open access to data generated through genetic studies has been most important in genetic research. In addition to the CEPH and Genographic databases, several public genetic databases exist, and many published DNA data are submitted to at least one of them. Most of the reputable journals require submission of DNA data to one of the major public, genetic sequence databases, such as the European Molecular Biology Laboratory (EMBL) database, hosted in Heidelberg, or the United States National Institute of Health (NIH) database, GenBank. We note that most anthropological and archaeological journals do not require submission of data to these databases, and we would encourage editors and reviewers of journals to recommend a change in policy. These databases are fully searchable and contain large amounts of DNA sequences and other genetic data, which increase daily. The complete genomes of several model organisms can now be found in GenBank and EMBL, and researchers are continuing to try to understand how those DNA sequences relate to genes, proteins, and the numerous phenotypic characteristics of the organisms as a whole.

In the same way that the development of PCR and automated sequencing dramatically increased the amount of DNA data generated in the late 1980s, the most recent technological development, known as next generation sequencing (NGS), may increase the scale once again by several orders of magnitude (Rothberg and Leamon 2008). While the production of the first complete human genome took thirteen years and cost billions of dollars, the latest development in NGS technology is revolutionizing genetic studies. We are rapidly approaching the point where an entire human genome can be sequenced for around US$1000 (Service 2006). This approach not only makes obtaining DNA sequences for modern samples faster and cheaper, it also has significant implications for aDNA studies. Next generation methods have allowed for the sequencing of the complete genomes of a Neanderthal (Green et al. 2010), as well as other ancient humans and hominins (discussed in chapter five). Such studies were not possible using the previous technology. Using NGS, researchers have been able to sequence parts of the genomes from archaeological horse remains, for example, to document changes in characteristics such as coat color through the process of domestication (Ludwig et al. 2009; see chapter seven for a discussion of this study). By the time this book goes into distribution, there will undoubtedly have been new publications applying the techniques to problems we have not yet considered.

Structure of This Book

This first chapter has discussed our perceptions about the need for this book and explained our perspectives and approaches in writing it. We have given a brief history of molecular anthropology and some of the developments in DNA technology that have led us to a point in which the worlds of archaeology and molecular biology can become intertwined. Throughout the rest of the book, we hope to provide the readers with a basic background and introduction to the terminology and concepts that are necessary to understand DNA research. We will also provide examples of the way that DNA has been and can be used for addressing issues of interest to archaeologists and others interested in history and prehistory.

The first half of this book focuses on the basics of DNA (chapter two) and aDNA (chapter three): what it is, how we analyze it, and an important issue that is often overlooked (Cann 1994), the ethics of dealing with DNA samples (chapter four). Those not interested in or already familiar with the basics of DNA may want to skip chapter two. While chapter three includes some basic information about what aDNA is, it also provides some key information regarding the collection of archaeological samples for aDNA studies and some of the precautions archaeologists should be familiar with when collecting samples they might want to use for aDNA research.

The second half of the book provides examples of the various contributions aDNA has made to studies of hominin evolution (chapter five) and the applications of both ancient and modern DNA research in archaeology, including: the study of population origins and migrations (chapter six); the advent of domestication and other human impacts on plants and animals (chapter seven); and health and demographic factors that influence individuals (chapter eight). We focus on case studies in each chapter rather than reviewing the entire literature on the subject. Throughout the book, we also highlight common misconceptions or frequently asked questions regarding DNA, molecular anthropology, and the application of genetic approaches to archaeological issues.

Chapter Two

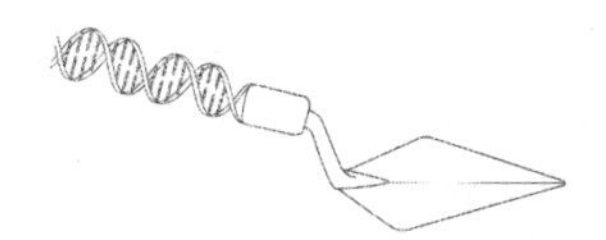

DNA: The Basics

Since Watson and Crick (1953a, 1953b) discovered the structure of DNA, researchers in many disciplines have exploited it. There are two major motivations for the archaeologist's interest in DNA. First, it is a molecule that encodes information about its own history within its structure, and as such, it is a "document of evolutionary history" (Zuckerkandl and Pauling 1965:357). Second, DNA analyses can illuminate important biological properties of organisms, such as their sex, species, and the changes in functionality of particular **genes** through time and across space. Some of these properties, such as age, sex, or species identification can be determined using traditional archaeological approaches; however, often they cannot. This is where DNA identification is particularly useful.

This chapter will address some of the fundamentals of DNA structure and function and describe some of the methods used in DNA analyses commonly applied to address archaeological or other anthropological issues. Like the rest of this book, it is not intended for someone wanting to engage in molecular research. It is intended as a primer for archaeologists interested in becoming critical consumers of published molecular data that address anthropological questions. It is also aimed at those who are interested in developing working relationships with molecular anthropologists and making the most of archaeologically derived genetic data.

DNA: What is It and What Does It Do?

DNA stands for deoxyribonucleic acid, which refers to the chemical structure of the molecule. If the reader would like a detailed definition

---

DNA for Archaeologists by Elizabeth Matisoo-Smith and K. Ann Horsburgh, 21–58.

of the term, biology textbooks are a good source. For orientation here, DNA is common to almost all organisms (some viruses, for example, do not contain DNA) and has two main roles—protein synthesis and self-replication. In addition, it serves to regulate both of these processes. That is, it provides the necessary information and instructions for the production and function of every cell and system in our bodies. In order for an organism to grow, develop, and repair itself, cells must replicate. The body is, of course, much more than just a mass of cells—it is an intricately orchestrated interaction of organs and systems. Thus, it also requires the provision of the instructions for the production of proteins, the substances that hold the body together and allow it to function.

Each human body is made up of tens of trillions of cells. These cells are classified into three different types: somatic, germ, and stem. In all these cell types, the majority of the DNA is found in the nucleus and organized in structures called **chromosomes**. This DNA is known as **nuclear DNA**, sometimes referred to as nDNA, as opposed to the DNA that exists outside of the cell nucleus in the mitochondria, which is known as mtDNA (which will be discussed later in this chapter). Humans have forty-six chromosomes in total, arranged in twenty-three pairs. One of each pair of chromosomes originates in each parent, such that there is a maternal and a paternal copy of each chromosome in each cell of the body (except for red blood cells, which do not have a nucleus) (Figure 2.1). Of the total forty-six chromosomes in each cell, two are the **sex chromosomes**, designated X and Y in mammals. The remaining forty-four chromosomes are called **autosomes**. Each pair is designated 1–22 and is numbered from the largest to the smallest in size. Humans normally have either two X chromosomes if they are female, or an X and a Y if they are a male. Occasionally, problems occur during **meiosis** (see the cell division discussion below) that may result in the inheritance of too many or too few chromosomes. One such case is the condition chromosome 21. Another, Turner's syndrome, are three copies of chromosome 21, and another, Turner's syndrome, also known as monosomy X, is when a female only has one X chromosome.

Chromosomes are simply super-coiled strings of DNA, and genes are the portions of DNA that encode for a particular protein or function. Researchers have now "mapped" many genes, such as those associated with particular disease risks, to their exact location on specific chromosomes. For example, we know that the gene associated with the disease cystic fibrosis is located on chromosome 7, and that the gene associated with sickle cell anemia is located on chromosome 11. The DNA itself is a sequence of **nucleotides** or bases, represented by the four letters, A, T, C, and G (Figure 2.2). The entirety of the DNA in a cell is

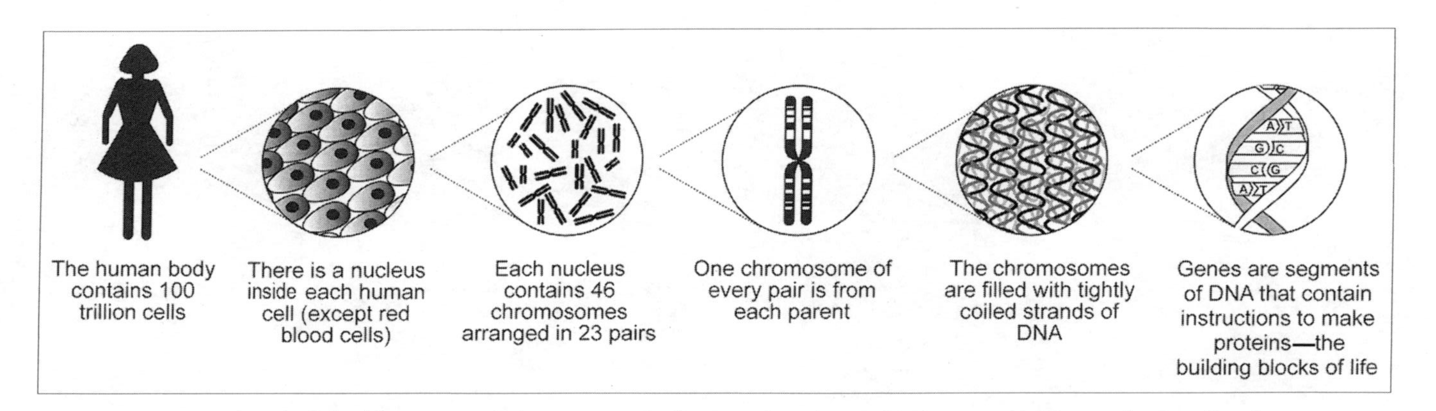

Figure 2.1 From the whole body to DNA—A lesson in scale showing the relationship between the human body, cells, chromosomes, genes, and DNA sequence.

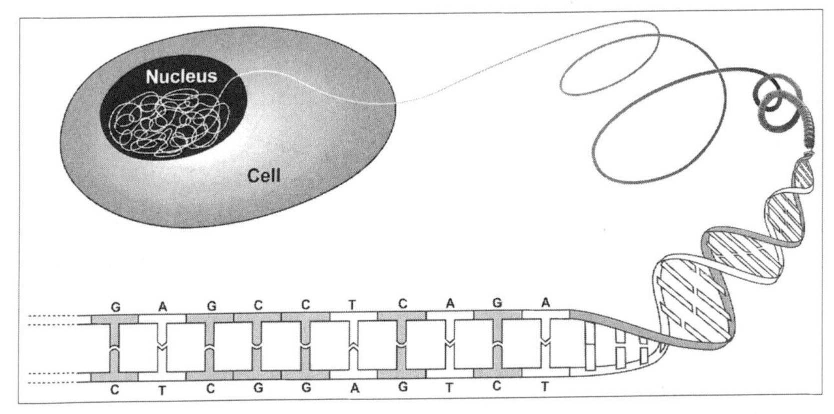

Figure 2.2 The nucleotide structure of DNA showing bases A, T, C, and G.

known as the **genome**, although often people will talk more specifically about the nuclear genome and/or the mitochondrial genome. The entire human genome was recently determined to be approximately 3.2 billion bases (also referred to as base pairs or bp) long (Human Genome Project 2012).

The majority of the cells in our bodies are **somatic cells**, and when people talk about cells, generally, they are referring to somatic cells. While somatic cells are differentiated by function, every somatic cell in the body contains the same DNA sequence; they just use or translate different parts of that DNA sequence for their specific function. In other words, skin cells, white blood cells, and brain cells all have exactly the same copy of DNA, but are differentiated by the expression of different genes in each of those cells.

Germ cells, also known as sex cells or **gametes** (egg and sperm), differ in that they contain half the amount of DNA found in somatic cells. This half-complement allows a fertilized egg, or **zygote**, to inherit genes from both parents without doubling the total amount of DNA in each generation. This process will be discussed later. It is through the germ cells that genetic information is passed from one generation to the next.

Stem cells, which have been the topic of much public debate, are cells that are defined by their ability to become any number of different somatic cell types. They are classified into two general types—embryonic and adult. Once a stem cell has taken the path to becoming a particular type of cell, a process known as cell differentiation, it cannot go back to being a stem cell—once a skin cell, always a skin cell. Adult stem cells are found in bone marrow and are important for the regeneration of specialized cells, such as blood cells. They are also found in the umbilical cord blood of newborn babies (yes, even though they come from a baby, they are still known as adult stem cells). Cord blood samples, which can be frozen and stored for future use, have become an important source of stem cells. Like the stem cells removed from a bone marrow donor, they are clinically utilized for repair and replenishing specialized cells that have been depleted, for example, through chemotherapy or other cancer treatments. It is hoped that ongoing research to harness the potential of stem cells for tissue regeneration will provide advances in the treatment of many conditions including brain or spinal cord injuries. The majority of public debate about the use of stem cells in research applies to the use of embryonic stem cells, which are obtained from four-to-five-day-old embryos grown in the laboratory, often for in vitro fertilization.

The Structure and Function of DNA

As stated earlier, one of the major functions of DNA is to coordinate its own replication. Its ability to do this is based largely on its physical and chemical structure. It is composed of the four bases (also referred to as nucleotides) adenine, cytosine, guanine, and thymine, which are abbreviated A, C, G, and T. The bases C and T are known as **pyrimidines**. And the other two, A and G, are known as **purines**, with these names referring to their particular chemical structure. The twisted ladder-like, or double helix, structure of DNA was determined in 1953 by James Watson, Francis Crick, Rosalind Franklin, and Maurice Wilkins (Franklin and Gosling 1953; Watson and Crick 1953a, 1953b; Wilkins et al. 1953) and is illustrated in Figure 2.2. There are strong chemical bonds along the outsides of the ladder, with weaker bonds across the rungs. These weaker bonds hold together the nucleotides on the two different sides of the ladder. The nucleotides can only bind in a particular way: A to T and C to G. Thus, the DNA molecule is made up of two complementary strands of DNA. Therefore, if you know the sequence of nucleotides on one strand, you can immediately reconstruct the sequence of the other strand. This is important for DNA replication. It is also why DNA researchers will talk about DNA sequences in terms of the number of base pairs that they used in their analyses, as in "250 bp of mtDNA were sequenced." Note that this can be somewhat confusing when DNA sequence data is combined with discussions of archaeological dates in which BP (before present) is often used for uncalibrated radiocarbon dates.

During DNA replication, which is the first step in the process of cell division, part of the DNA molecule unwinds, and the weaker bonds on the rungs of the ladder break—in a process similar to unzipping a zipper. Once the ladder is open, free flowing nucleotides (As, Ts, Cs, and Gs) that exist in the nucleus of the cell are attracted to the exposed nucleotides still attached on either side of the rungs of the ladder. The attraction is always to the complementary base, such that any exposed A will attract a T, and any G will attract a C, and vice versa. This process occurs on both halves of the unzipped molecule, and thus, two new identical DNA molecules are constructed from the original. The cell can then divide producing two daughter cells, each with an identical copy of the entire genome. In mitochondria, the DNA is replicated independently of cell division. When the cell divides, approximately half of the total number of mitochondria segregate into each daughter cell.

Cell Division

We have been told that it was the endless drudgery of lessons on meiosis and **mitosis** in high school biology classes that contributed to many of our colleagues choices to study archaeology, rather than the biological sciences, when they left high school. So why should you care about it now? In order to understand why molecular anthropologists and biologists study particular regions of the genome, such as mtDNA or Y chromosomes, one must understand why they are different from other bits of DNA—and this is partly due to the processes that occur during cell division. The process of meiosis is also important for understanding patterns of genetic inheritance, which is after all, the key to understanding the historical events that archaeologists trying to reconstruct.

Cell division is all about making new cells, and broadly speaking there are two types of cells that need to be made. First, we will talk about the process of making somatic cells which constitute the tissues of your body. Then we will discuss the production of gametes, or sex cells, which are central to reproduction.

The process of cell division that produces somatic cells (and stem cells) is known as mitosis. The important feature of mitosis is that the entire genetic complement of the cell is duplicated to produce one copy for each daughter cell, generating two identical daughter cells in each division. Sexual reproduction requires meiosis, or the production of sex cells (eggs and sperm), which is a bit more complicated (Figure 2.3). In organisms that reproduce asexually, things are relatively simple; the DNA is replicated, the cell splits in two, and the resulting daughter cells are identical to the starting cell. In sexual reproduction, though, the genomes of two individuals need to be blended while preserving the same total number of chromosomes. This means that sex cells can only have half the number of chromosomes, while maintaining one copy of each chromosome type.

Meiosis is a two-step process, the first of which is DNA replication and cell division much like mitosis. Each of the two daughter cells

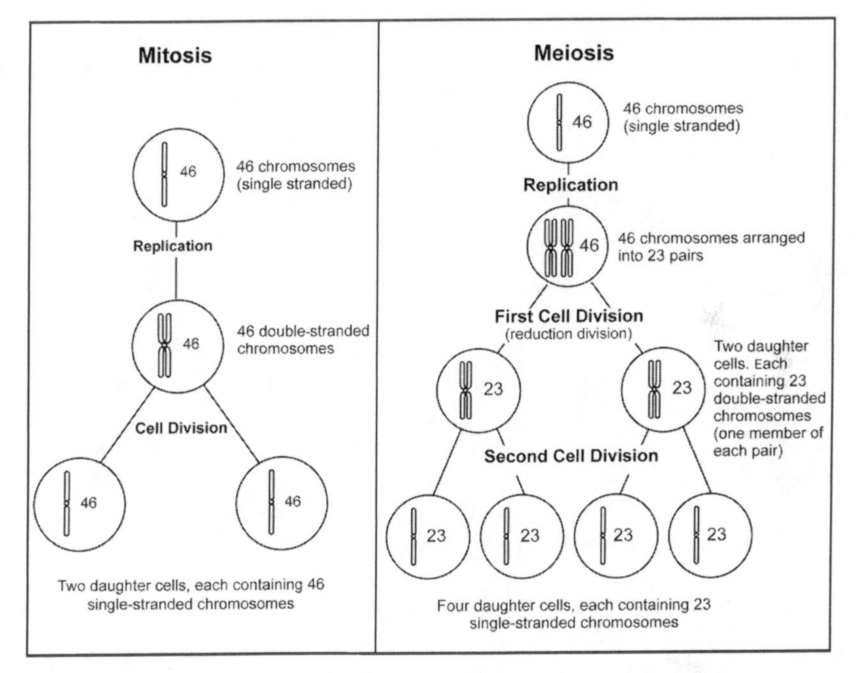

Figure 2.3 Representation of cell division during mitosis and meiosis.

then divide again without further replicating the DNA, producing four daughter cells, each with only one copy of each chromosome. This is also known as **reduction division**, as the total number of chromosomes is reduced by half during the process. Reduction division is essential in sexual reproduction because without it the fusion of human egg and sperm would result in the next generation having ninety-two chromosomes. Chromosomal number is important. Having too many or too few chromosomes is detrimental because of loss or duplication of important genetic information. This generally results in a nonviable zygote. As discussed earlier, some chromosomal variations are viable but generally result in developmental disabilities in the offspring, such as in the case of Down or Turner's Syndromes. The fusion of two chromosomes does occasionally occur; and as long as the information in the genes on those chromosomes is not lost or shuffled during this fusion, the resulting offspring could function normally. Evidence suggests that this probably happened at some point early in hominin evolution because humans have only twenty-three pairs of chromosomes, where as chimpanzees and the other great apes have twenty-four pairs (Ijdo et al. 1991).

If the DNA duplication and reduction process of meiosis were really this simple, an individual would inherit one copy of chromosome number three from his or her mother and one from his or her father; and the one inherited from the mother would have been identical to that of either *her* mother or father. Meiosis is not, however, that simple. A process called **recombination** also happens. During meiosis, each pair of chromosomes lines up and proceeds to swap sections of DNA before the cell divides. When a woman's eggs were being created, the genes she inherited from her mother were shuffled with the genes she inherited from her father, creating a completely new combination of DNA in each daughter cell. Every egg carries a different combination of a woman's inherited DNA. Similarly, each sperm produced by a man would carry a unique combination of his DNA. As a result, each time that fertilization occurs, it brings together a unique combination of DNA from each parent. Thus, while you may look similar to any siblings you might have, you do not look identical to them unless you have an identical twin. Recombination occurs along all the chromosomes from 1 to 22. However, the sex chromosomes recombine in only a small region, known as the **pseudoautosomal region**, but do not recombine over the majority of their length.

What Does DNA Do?

In addition to self-replication, the other key function of DNA is the encoding of information for building proteins. They are actually the second largest component of our body weight after water, and it is often said that proteins are the building blocks of organisms. Proteins are simply chains of units called amino acids. The DNA sequence in a gene determines the sequence of amino acids in a protein, which then determines the character and function of the protein. Proteins are commonly thought of as structural, such as in the collagen that is found in your bones or the lips of Hollywood starlets. They are also important for other bodily functions. Haemoglobin, for example, which transports oxygen around the body is a protein, as are the many enzymes key in their role as catalysts in numerous biological functions.

It is the sequence of the subunits (also known as nucleotides or bases) in the DNA that carries the information for protein synthesis, and those nucleotides are "read" like letters in a word. The order matters. In DNA, each word, or codon, is three bases, or letters, long. Each codon specifies and attracts a particular amino acid. Proteins are merely strings of amino acids assembled in different ways. The functional information in a gene is flanked by start and stop codons that designate the parts of the DNA sequence that are to be read and, thus, translated into a protein. Given that there are four bases, this means that there are 64 (4^3) possible three-base codon combinations. There are only twenty amino acids that make up proteins, so most amino acids are specified by more than one codon (Figure 2.4).

		2nd base							
		T		C		A		G	
1st base	T	TTT	(Phe/F) Phenylalanine	TCT	(Ser/S) Serine	TAT	(Tyr/Y) Tyrosine	TGT	(Cys/C) Cysteine
		TTC	(Phe/F) Phenylalanine	TCC	(Ser/S) Serine	TAC	(Tyr/Y) Tyrosine	TGC	(Cys/C) Cysteine
		TTA	(Leu/L) Leucine	TCA	(Ser/S) Serine	TAA	*(Stop)*	TGA	*(Stop)*
		TTG	(Leu/L) Leucine	TCG	(Ser/S) Serine	TAG	*(Stop)*	TGG	(Trp/W) Tryptophan
	C	CTT	(Leu/L) Leucine	CCT	(Pro/P) Proline	CAT	(His/H) Histidine	CGT	(Arg/R) Arginine
		CTC	(Leu/L) Leucine	CCC	(Pro/P) Proline	CAC	(His/H) Histidine	CGC	(Arg/R) Arginine
		CTA	(Leu/L) Leucine	CCA	(Pro/P) Proline	CAA	(Gln/Q) Glutamine	CGA	(Arg/R) Arginine
		CTG	(Leu/L) Leucine	CCG	(Pro/P) Proline	CAG	(Gln/Q) Glutamine	CGG	(Arg/R) Arginine
	A	ATT	(Ile/I) Isoleucine	ACT	(Thr/T) Threonine	AAT	(Asn/N) Asparagine	AGT	(Ser/S) Serine
		ATC	(Ile/I) Isoleucine	ACC	(Thr/T) Threonine	AAC	(Asn/N) Asparagine	AGC	(Ser/S) Serine
		ATA	(Ile/I) Isoleucine	ACA	(Thr/T) Threonine	AAA	(Lys/K) Lysine	AGA	(Arg/R) Arginine
		ATG[A]	(Met/M) Methionine	ACG	(Thr/T) Threonine	AAG	(Lys/K) Lysine	AGG	(Arg/R) Arginine
	G	GTT	(Val/V) Valine	GCT	(Ala/A) Alanine	GAT	(Asp/D) Aspartic acid	GGT	(Gly/G) Glycine
		GTC	(Val/V) Valine	GCC	(Ala/A) Alanine	GAC	(Asp/D) Aspartic acid	GGC	(Gly/G) Glycine
		GTA	(Val/V) Valine	GCA	(Ala/A) Alanine	GAA	(Glu/E) Glutamic acid	GGA	(Gly/G) Glycine
		GTG	(Val/V) Valine	GCG	(Ala/A) Alanine	GAG	(Glu/E) Glutamic acid	GGG	(Gly/G) Glycine

Figure 2.4 The amino acid table showing the three-base codons and their corresponding amino acids with the standard abbreviations and stop codons. The codon ATG (Methionine) is the start codon.

For example, the amino acid glutamine (abbreviated Gln or Q in molecular biology shorthand) is designated by the two DNA codons CAA and CAG. The most important letters in the three letter codon are the first two. The third codon position can often vary without changing the identity of the amino acid. This redundancy in the coding for amino acids (both in the third codon position and by the fact that most amino acids have multiple codons) means that much variation in the DNA sequence will have little or no functional impact.

Despite the fact that there are at least 25,000 known genes in the human genome, each of which is on average 3,000 bp long, the majority of the DNA in the human genome does not code for proteins and is often referred to as noncoding. While much of this noncoding portion of the genome is colloquially referred to as "junk," at least some of it has known functions in the maintenance of chromosome structure. Further, some of it is involved in the regulation of protein production, controlling when and where a particular protein is synthesized. Although some of the apparently nonfunctional DNA likely has some function, we have not yet learned what that might be. It does seem, however, that there are large sections of DNA that may have had a function in the past but are now nonfunctional (but not harmful) and are, in a sense, just along for the ride. Studying variation in these nonfunctional and other noncoding regions of the genome is particularly valuable for reconstructing population history and migration because their presence in a population is more likely to be due to ancestry rather than to natural selection.

Mutations

How does the variation that is so useful for reconstructing population history occur? Identifying the origin of variation both within and between species was the major problem with which Darwin grappled in developing his theory of natural selection. It was not until forty years after the publication of *On the Origin of Species* that the main source of variation, **mutation**, was discovered by the Dutch botanist, Hugo de Vries. In the late 1890s, de Vries was conducting studies of inheritance in primroses when he found that new varieties would spontaneously appear. Once these new traits, which de Vries called "mutations" appeared, and the plants were crossed with the standard varieties, the new traits were inherited in future generations as predicted. In addition to discovering that mutations were the source of variation, de Vries also "rediscovered" the laws of inheritance that Gregor Mendel identified in his now famous studies of pea plants. Interestingly, de Vries' maternal grandfather, Caspar Reuvens, was the world's first professor of archaeology, appointed at Leiden University in 1818 by King William I, the first king of the Netherlands.

Box 2.1: Mutations

Mutations can be individual changes in a base, like a change from an A to a C (referred to as an SNP and pronounced snip), or they can be insertions or deletions (indels) of bases. Mutations are more likely to accumulate in regions that do not code for any protein because any change in the coding regions will alter the instructions for the construction of the organism. A major change in protein production is likely to be detrimental and cause the death of the organism, which will result in those genes not passing to the next generation. As a result, mutations are more likely to accumulate in portions of the genome where they will have no functional implications. Obviously, mutations can occur in functional regions of the genome, and as long as these mutations do not severely impact the viability of the organism, they will be maintained in the population.

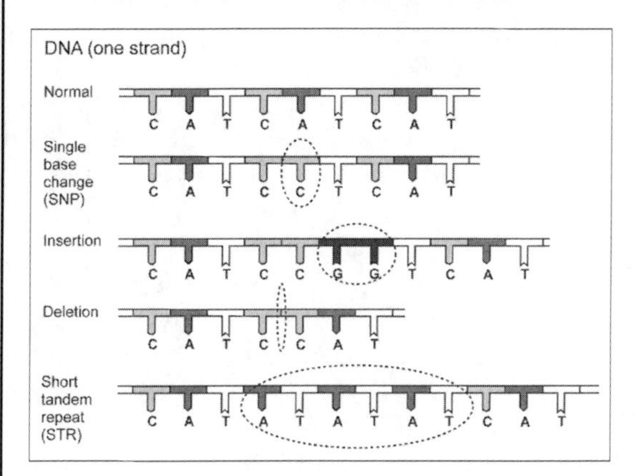

In addition to occurring more frequently in particular regions of the genome, different types of mutations are more common than others. Among SNPs, mutations are more likely to occur between bases that are structurally similar. The structures of cytosine and thymine are similar, and they are known as pyrimidines. The structures of adenine and guanine, which are known as purines, are also similar. Therefore, we see a lot more pyrimidine-to-pyrimidine or purine-to-purine mutations (C↔T or A↔G), which are called transitions, than purine to pyrimidine mutations, known as transversions.

Insertions or deletions can be of single or multiple nucleotides. A common form of mutation can be a "stutter" during DNA replication, producing repetitions of relatively short sequences. For example, a nucleotide pair of AT can be repeated multiple times. The number of bases in these repeat units can be small (1–8 bp), which are known as STRs and are often referred to as microsatellites. Mutations of larger repeat units (9–100 bp) also occur, and these variable regions are called minisatellites. Some of these mini and microsatellites lie within coding regions, and therefore, they affect gene products. Many are found within noncoding regions, which leaves them selectively neutral and unconstrained by natural selection. Minisatellites tend to mutate more frequently than SNPs, and therefore, they can provide more information about variation within populations over short periods of time. Furthermore, they tend to mutate in a stepwise fashion such that they increase or decrease in size by single-unit increments. The rapid rate of mutation of STRs, however, means that two individuals could be carrying the same STR repeat number for a particular marker, but the similarity may be the result of two mutation events from nonidentical, ancestral sequences meaning they do not have a shared recent ancestry. Their high levels of variability mean that STRs can be very useful for forensic or paternity testing when matching DNA profiles of specified individuals are necessary. However, they are only powerful when many STR markers are used in conjunction and when investigating evolutionary relationships must be combined with more slowly mutating markers.

Every so often an error is made in the copying of the DNA sequence—like a typo. That error is a mutation. In addition to errors caused by mistakes in the copying of the DNA, mutations can be caused by environmental insults such as UV rays or chemicals. Generally, if a mutation occurs in a somatic cell and is not immediately repaired by DNA-repair enzymes, it will be faithfully replicated and found in all the daughter cells of the originally mutated cell. Somatic cell mutations will not be passed on to future generations. This is what happens, for example, in tumor growth. Not all mutations, however, have functional implications and can simply be unexpressed. These are often referred to as silent mutations. In order for a mutation to be inherited, it must occur in a sex cell.

Mutation is the main source of variation upon which natural selection can act. One of the most important concepts regarding mutations and one that is misunderstood by many when they discuss natural selection, is that mutations occur randomly. Organisms do not acquire mutations based on what they need. If a mutation occurs and happens to affect the reproductive success of the individual, then it will become either more or less frequent in future generations depending on whether it has a positive or a negative effect on the reproductive success (fitness). If a particular genetic marker or mutation is variable in a population, then it is referred to as being polymorphic (meaning many morphs or types) or a **polymorphism.**

Most mutations that occur and are inherited, however, do not affect the organism in any way. This is because they often occur in the noncoding regions of the DNA or in the third codon position, as described above. Once a mutation occurs in a sex cell, though, it can and will be passed down from generation to generation—particularly if it is a silent mutation. Although mutations are random, they do tend occur regularly, and if not subject to natural selection, these mutations accumulate in a clock-like fashion. We refer to the frequency of mutations as the mutation rate, which is generally expressed in terms of the number of mutations per generation. We can use these mutations as markers to track change through time. When we compare or align two or more DNA sequences, we can reliably assume that individuals or species that have more similar DNA shared a common ancestor more recently than those with dissimilar DNA. By comparing the similarities and differences in the DNA sequences of many individuals, we can reconstruct a kind of family tree or a phylogeny. The mechanics of mutations are discussed in detail below under methods, and the reader is referred to Box 2.1 for a visual representation.

Mitochondrial DNA, Y Chromosomes, and Other Useful Markers

Most of the molecular anthropological work of interest to archaeologists addresses the origins and population histories of humans and their associated animals and plants. Given the complexity of the recombination that occurs in the chromosomes, combined with the effects of natural selection, reconstructing the history of most nuclear DNA variation is extremely difficult. Each generation is unique, and specific genes from the parents may or may not be inherited by the offspring. Fortunately, there are portions of DNA that can be inherited from only one parent. These bits of DNA still accumulate mutations, but they do not undergo recombination. This makes tracking the sequential mutations, reconstructing their histories, and obtaining sex-specific historical information much more simple. The two types of DNA inherited exclusively uniparentally are the mtDNA, which as referenced above, is only inherited through the maternal line (Birky 1995), and the DNA found on the **nonrecombining portion** of the Y chromosome (the NRY), which is only passed down from fathers to sons (Jobling and Tyler-Smith 1995).

Mitochondrial DNA

Mitochondria are small organelles that exist outside of the nucleus of the cell. They provide the energy necessary for cell function and are involved in cell respiration. There are, on average, some 2,000 mitochondria in every cell, and each mitochondrion contains several copies of its own genome, or mtDNA. As in nuclear DNA, mtDNA is made up of the bases A, T, C, and G, and virtually all of the mitochondria in the body carry the same copy of mtDNA. Unlike the DNA in the nucleus, however, the mtDNA is not bound in chromosomes but is a small circular molecule. In most multicellular organisms, the mitochondrial genome encodes thirty-seven genes and carries a noncoding "control region" of approximately 1,100 bp or 1.1 kilobase (1.1 kb) long. In humans, the entire mitochondrial genome is about 16.5 kilobases long (Ingman and Gyllensten 2001).

There are several qualities of mtDNA that make it particularly valuable for anthropological studies. The first, as discussed above, is the nature of its inheritance. Mitochondria, and therefore, mitochondrial genomes are almost universally maternally inherited. While males do inherit mtDNA, they do not pass their mtDNA on to their offspring. This is because the mitochondria in the sperm are few in number and do not enter the egg during fertilization, or if they do, they are generally rejected from the ovum.[1] Thus the sperm introduce only the paternal nuclear DNA to the fertilized egg. This means that barring the generation of a new mutation, offspring inherit an exact copy of their mother's mtDNA.

The second important quality of mtDNA for reconstructing population history is that it has a mutation rate approximately ten times higher than that of nuclear DNA (Brown et al. 1979). This is primarily because of a lack of efficient DNA-repair enzymes in mitochondria (Lansman and Clayton 1975). Despite the lack of repair enzymes, the genes in the mitochondrial genome play central roles in cellular metabolism. They are also tightly controlled by natural selection because nonsilent mutations, or those mutations that will cause an amino acid change, are likely to cause catastrophic loss of function. For this reason, much of the focus on mtDNA variation in modern human populations and studies of within-species variation in other organisms has focused on the noncoding control region—also sometimes referred to as the displacement-loop or the D-loop region—of the mitochondrial genome. In particular, they have focused on the parts of the control region that are **hypervariable.** Two of these regions have been designated hypervariable regions 1 and 2 (HVR 1 and HVR 2).

While rapid mutation rates are generally useful for reconstructing the ancestral relationships between closely related organisms, some parts of the mitochondrial genome are known to mutate so rapidly that the particular identity of an ancestral state is difficult to determine. These so-called "hotspots" are, therefore, phylogenetically uninformative, so they are often ignored in comparative studies. Such hotspots, however, are proving to be useful in that they are allowing us, in a sense, to "see" evolution in action. When sequencing the mtDNA of some individuals, researchers occasionally find that the mtDNA from a single individual shows some variation. For example, occasionally a sequence would have one base, say a T, at position 16342 in one cell but a C at that same position in another cell. This condition, known as **heteroplasmy**, is common in the mtDNA of tissues and is more common in older individuals because mutations and errors in copying are more likely to occur as cells age. However, when we see the same heteroplasmic site in population studies, this can be evidence of the creation of a new mtDNA lineage. Additionally, such heteroplasmic lineages are believed to change relatively quickly, and these sites can be markers of recently shared ancestry, as was demonstrated in the use of a heteroplasmic site to confirm the identity of the remains of Czar Nicholas II of Russia (Ivanov et al. 1996; discussed further in chapter eight).

In addition to its maternal inheritance and rapid mutation rate, the third factor making mtDNA particularly useful, especially for aDNA studies, is the quantity of mtDNA found in each cell. Where each cell contains one copy of the nuclear genome, that same cell contains thousands of copies of mtDNA. By virtue of its abundance, mtDNA is easier to analyze. Most importantly for archaeologists, it is more likely to be preserved in archaeological remains (see chapter three).

As discussed in chapter one, the first worldwide survey of human mtDNA variation was undertaken in the Wilson laboratory at the University of California, Berkeley, the results of which were published in 1987 by Rebecca Cann, Mark Stoneking, and Allan Wilson in their paper "Mitochondrial DNA and Human Evolution." Since this publication, methods of both DNA extraction and analysis have resulted in the generation of ever better mtDNA data. In addition, numerous researchers, and even major research consortiums (e.g., National Geographic's Genographic Project), have concentrated their efforts on providing better coverage of worldwide mitochondrial variation. These studies will be discussed more fully in chapter six.

The first complete human mitochondrial genome was sequenced in 1981 by researchers at Cambridge University. This genome is used as a reference sequence (Anderson et al. 1981), often referred to as the **Cambridge Reference Sequence** (CRS). Small sequencing errors in the original published version of the CRS were corrected and revised in 1999 (Andrews et al. 1999). Now sometimes known as the **rCRS**, it is against this sequence that all other published mitochondrial genomes are compared. Each base in the reference sequence is given a number that allows researchers to locate and refer to particular mutations they find. In standard notation, a mutation at position 16294 in which the rCRS has a C and a new **haplotype** has a T would be written C16294T. By comparing DNA sequences obtained from a large number of individuals, we find that many people will share the same mtDNA sequence and can be grouped together. Many of these groups will share common mutations while possessing others that are unique to their group. Computer programs are generally used to compare all of the DNA sequences in a study, identify those that are identical, and then compare those sequences with large international databases of other DNA sequences. This effort allows for the reconstruction of the evolutionary relationships between large numbers of individuals and groups of individuals.

After some twenty-five years of research into human mtDNA variation, we now have a relatively complete maternal phylogeny or family tree. While new mutations are inevitably going to be identified as more and more individuals are studied, the general structure of the tree is unlikely to change. Figure 2.5 shows the basic structure of the human mtDNA tree. Key mutations or combinations of mutations identify each major lineage or branch of the tree, which has been assigned a letter. For example, the sequences that are the most different from all of the others form the deepest branches on the tree. These have been dubbed "L" lineages. The L3 branch splits into two branches, dubbed M and N, which then split into new branches, each of which are given a letter.

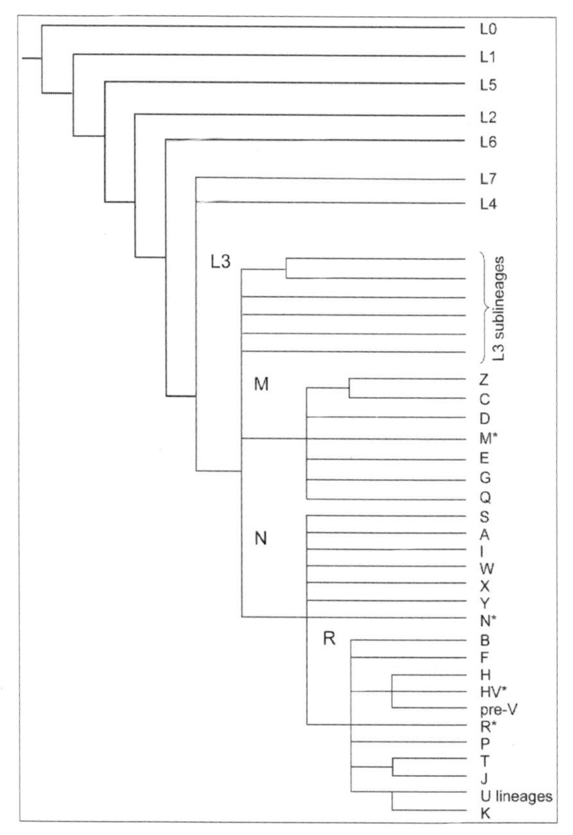

Figure 2.5 The basic structure of the human mtDNA phylogeny.

When those branches are further divided into subbranches, each new subbranch is given a number (e.g., B1, B2, B3, and B4). Further subdivisions of those subbranches are given another letter suffix (e.g., B4a, B4b, and so on). Thus, each time a new subbranch or new mutation on a lineage is found, it is given a distinct letter or number designation, depending on where it fits in the human mtDNA tree (Figure 2.6).

The DNA sequences at the ends of the branches on a phylogenetic tree, or those letters that relate to a DNA sequence with a specific combination of mutations, are referred to as haplotypes. Groups of haplotypes that share a common ancestor and, thus, share a number of specific defining mutations are said to belong to a **haplogroup** (also sometimes referred to as a **clade**), which has an older origin than the mutations defining the haplotype. So it can be assumed that the shared ancestry (sometimes referred to as the most recent common ancestor, MRCA) of haplogroup H is older than the shared ancestor of haplogroup H1. Note,

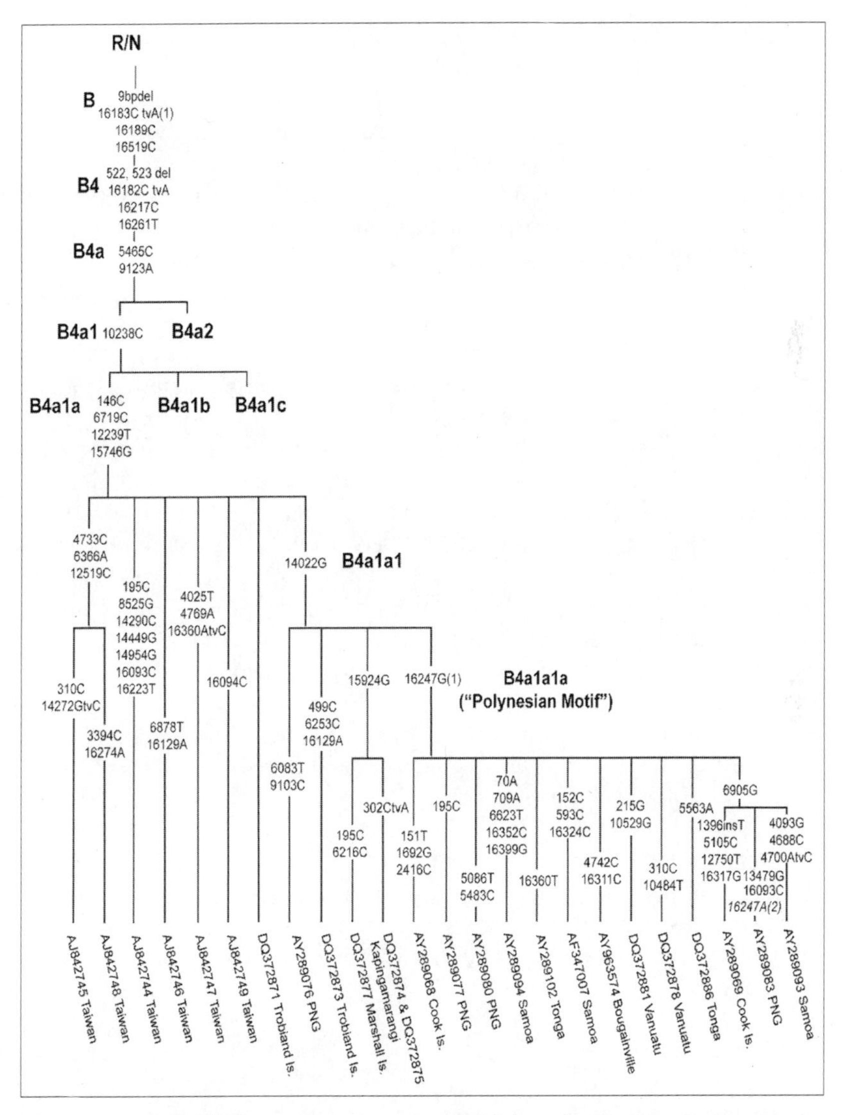

Figure 2.6 A detailed representation of mtDNA from haplogroup B showing the defining mutations for each subbranch of the B4 clade, including B4a1a1a which is sometimes known as the Polynesian motif. Reproduced with permission, after M. Pierson (2007), unpublished PhD thesis, Canterbury University.

though, that while the mutations defining the H1 haplogroup may be ancestral to the H1a haplotype, this does not necessarily imply that an individual with an H1 haplotype is the ancestor of an individual with an H1a haplotype. The phylogenetic history of a haplotype is often written as a nested hierarchy; for example, haplogroup P, which is a subtype of haplogroup R, which is itself a subtype of macrohaplogroup N, can be written N:R:P.

While the early mtDNA studies concentrated their analyses on characterizing variation in the noncoding control region, it is now recognized that mutations in the coding regions of the mitochondrial genome also carry significant phylogenetic signals. As such, there is an increasing move toward sequencing complete mitochondrial genomes. In addition to being more informative, it is also technically more accessible because of the development of **next generation sequencing** (NGS) technology, which makes whole-genome sequencing economically practical (Millar et al. 2008). Perhaps more importantly, complete mitochondrial genome sequencing alleviates the problem of **ascertainment bias.**

Ascertainment bias occurs when conclusions about populations are drawn from inadequate or nonrepresentative data. Most of the early evidence we had about biological variation came from studying particular populations. We were, therefore, unaware of variation in underrepresented populations. Research is frequently targeted to survey patterns of biological variation that we already know exist. Consequently, existing but unexpected variation may not be observed because it is not looked for. This was a particular problem with early mtDNA studies and those that only dealt with the control region of the mitochondrial genome. Problems deriving from unexpected variation, however, are alleviated by sequencing whole-mitochondrial genomes (Richards and Macaulay 2001). While complete mitochondrial sequencing is helping to address ascertainment bias, the almost exclusive focus on mtDNA throughout the 1980s and 1990s introduced another source of bias to studies of population dispersal or expansion. By reconstructing human migration history based only on variation passed down through the female line, we were only studying the history of half of the human population.

Y Chromosomes

It is the possession of a Y chromosome that determines that an individual is male—women do not have a Y chromosome—so it must be inherited by males from their father. Because the Y chromosome is found in the nucleus, it behaves like the other chromosomes and lines up with its homologous X chromosome just before cell division. During this process, a specific part of the Y chromosome, known as the pseudoautosomal region, can recombine with the X chromosome, but the majority of the Y chromosome does not. Generally, when we refer to the Y chromosome in evolutionary studies, we are talking about the nonrecombining portion (NRY), which makes up approximately 95% of the Y chromosome (Jobling and Tyler-Smith 1995).

While the first major Y chromosome variant or polymorphism was discovered in 1985, it was not until ten years later that researchers had enough data to start to address worldwide variation and compare those patterns with those obtained from the mtDNA studies. The Y chromosome is paternally inherited, so it provides a compliment to the mtDNA data for reconstructing population origins and migrations by tracing the movement of males rather than females. Interestingly, in many cases, mtDNA and Y chromosome studies show quite different patterns of population relationships. For example, in the Pacific, studies show that most Polynesian mtDNA lineages can be traced back to an Asian origin, with a limited number of them originating in Near Oceania. On the other hand, Y chromosome studies indicate much more of a Near Oceanic paternal contribution to Polynesian populations (Kayser et al. 2000).

The most likely cause for the different patterns in mtDNA and Y chromosome variation in any given population is that the mobility of men and women are likely to be different depending on kinship practices and corresponding regulations regarding whether the man or the woman moves when a couple marries. In the Pacific, the different patterns of mtDNA and Y chromosome diversity, it has been argued, may be explained by the matrilineal and matrilocal nature of many Pacific societies (Hage and Marck 2003). Part of this difference in mtDNA and Y chromosome patterns is also likely due to differences in the mutation rate. Interestingly, despite the relatively high mutation rate in mtDNA, the NRY exhibits an even higher rate (though only in certain kinds of mutations as discussed below). This is, in part, due to the high number of cell divisions Y chromosomes experience during spermatogenesis, the production of sperm cells (Jobling et al. 2004). A third factor influencing the geographical patterning of Y chromosome and mtDNA diversity is demography because a man can father more than one hundred children in his lifetime, whereas women can only give birth to a few. Further, males exhibit higher early life mortality than females; and there is greater frequency worldwide of polygyny than polyandry.

In the fifteen years after the discovery of Y chromosome polymorphisms, and the recognition that they could provide valuable data for reconstructing the dispersal of males through time, many labs around the world began to study Y chromosome diversity. By 2001, there were at least seven different nomenclature systems for classifying Y chromosome diversity, making interlab comparisons of data difficult. Unlike mtDNA, which is only 16.5 kilobases long, the Y chromosome is approximately 60 million bases long (Hammer and Zegura 2002), far too much for most labs to sequence. As a result, Y chromosome diversity is generally surveyed for known combinations of mutations, such as single nucleotide polymorphism (SNPs), different types of repeat sequences, and indels (also known as insertion or deletion mutations or binary polymorphisms in the Y chromosome literature; see Box 2.1 for the discussion of different mutation types). This results in potentially introducing significant ascertainment bias, which as discussed above, in terms of mtDNA studies, is now being overcome through studies of complete mitochondrial genomes.

In an attempt to deal with problems of ascertainment bias in Y chromosome studies, geneticists have identified a number of specific sets of SNP and other binary markers to identify the basic structure of the Y chromosome phylogeny. Because the total number of the SNPs used in these panels remains relatively small, a significant proportion of diversity within any given haplogroup will inevitably be missed. Thus, the combined use of SNPs and another type of mutation known as short tandem repeats (STRs), or mutations that result in differing numbers of repeat sequences, was instigated to try to identify intrahaplogroup variation. The mutation rates of STRs are much higher than among SNPs on the Y chromosome, so data from these rapidly mutating portions of the NRY can be combined with the more slowly mutating SNPs to provide a view of the variation within populations.

In 2002, the Y Chromosome Consortium (YCC) was established in an attempt to bring order and consensus to Y chromosome diversity studies. According to the YCC (Ellis et al. 2002), the term haplogroup is applied to lineages defined by specific SNPs or indels. The term haplotype is used for the sublineages defined by variation in STRs on the NRY. Unlike the marker nomenclature used in mtDNA studies, the identification of Y mutations does not relate to any nucleotide position on the chromosome. It is merely a sequential marker from the lab that identified the SNP mutation. Initially, two main laboratories were involved in Y chromosome diversity studies; one was at Stanford University and the other at the University of Arizona. These labs used different nomenclature for labeling markers, by designating them as either M or P (e.g., M112 or P52), where M stands for *mutation*, and P stands for *polymorphism*. Many more labs are now adding to our understanding of Y chromosome diversity, and new markers with different prefixes are being added to standard screening panels. In addition to these numbered SNP markers, important binary markers such as the specific insertion mutation referred to as YAP, and several markers on the sex-determining region on the Y (SRY) are being used to define the structure of the Y chromosome tree. These STR markers are generally identified by the prefix DYS, which stands for DNA Y chromosome segment, followed by a number, such as DYS 458.

In the Y chromosome tree (Figure 2.7), the oldest lineages, which are found exclusively in Africa, are found on branches A and B. The

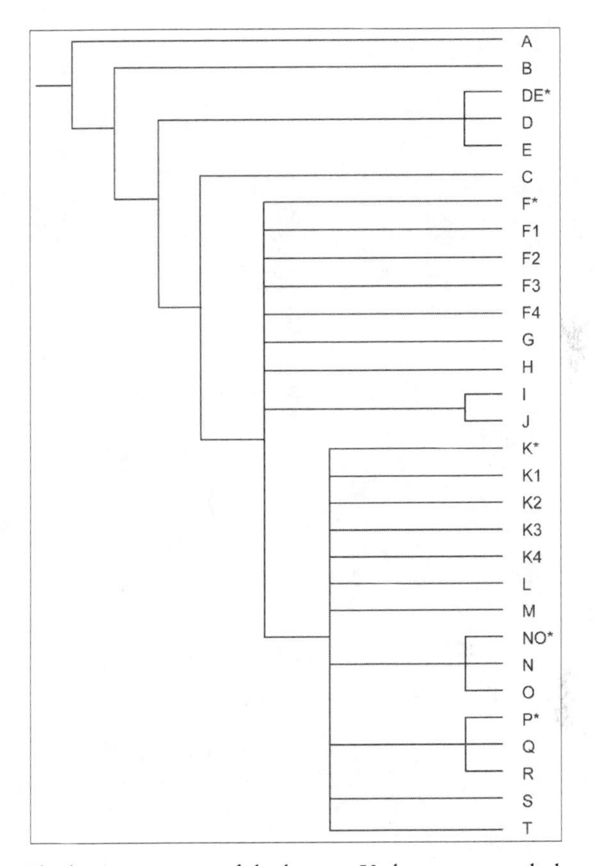

Figure 2.7 The basic structure of the human Y chromosome phylogeny.

YCC initially defined eighteen haplogroups (A–R), but, now (Karafet et al. 2008) twenty defined haplogroups are recognized (A–T). The sub-branches on the tree that possess the lineage-defining (or ancestral) SNP, but which do not possess the downstream, derived mutations are labeled with an asterisk (e.g., *K** is pronounced "*K star*"). Furthermore, among these particular haplotypes, researchers can show that they have screened an individual for a particular marker but that the polymorphism or SNP was not present. They denote this by showing the tested SNP or SNPs in parentheses. For example, O-M175* (xM119, M122, M95) means that this individual belongs to haplogroup O, carries the ancestral O defining SNP, M175, and has been screened for, but does not carry the derived M119, M122, or M95 SNPs (Figure 2.8).

Despite the YCC's best efforts to simplify the Y chromosome tree, two complementary systems are being used to define lineages. One uses an

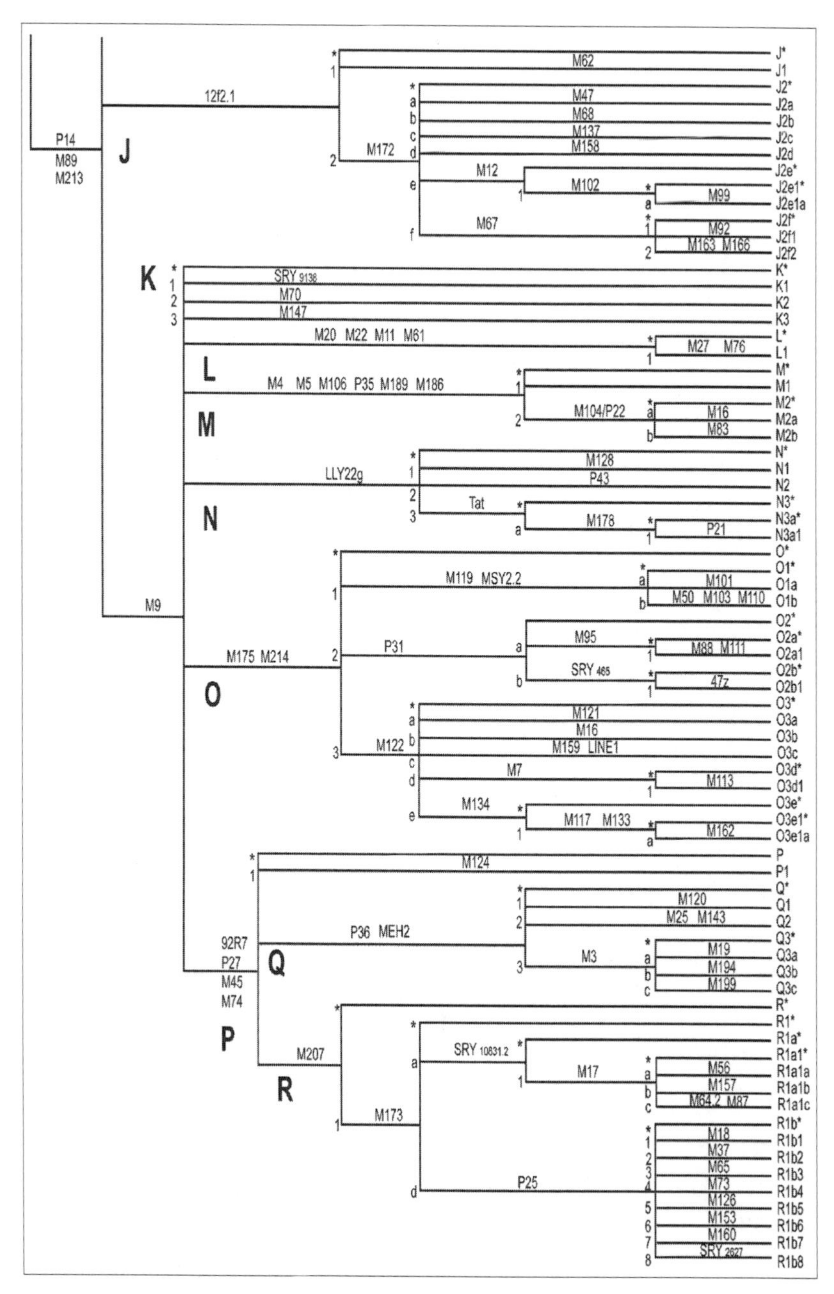

Figure 2.8 A detailed representation of half of the Y chromosome phylogeny defining mutations for each branch and subbranch (after Jobling et al. 2004:259).

alphanumeric system similar to the mtDNA system, such as R1b1, and the other identifies the lineages by the terminal mutation that defines them (e.g., N3-P21, or R1-M18). For the sake clarity, we look forward to the application of NGS to the widespread sequencing of the entire NRY. This will perhaps identify all or most of the phylogenetically informative markers of variation on the Y chromosome, which will allow for a more robust and evolutionarily informative phylogeny. Until that time, however, we are reliant upon this complex system. Therefore, our advice to archaeologists is: 1) be aware that the nomenclature has changed over time, so check the date of any references you are using; 2) make a list of common markers in the region that you study; and 3) consult with your collaborators at each step.[2]

Now that there are reasonable regional data sets for both mtDNA and the Y chromosome, some interesting implications of differences between the gene trees are being considered. Social structures such a matrilocality and patrilocality and their impacts on genetic structure have been investigated (Hage and Marck 2003). Similarly, differing patterns of interaction suggesting population settlement vs. invasion have been indicated by studies in Britain and Iceland (Als et al. 2006; Goodacre et al. 2005; Helgason and Stefánsson 2003; Helgason et al. 2000; Helgason et al. 2001; Helgason et al. 2009; Melchior et al. 2008). Additionally, there are interesting patterns of the geographic spread of specific Y chromosomes that appear to be associated with particular male-dominated, militaristic or religious expansions, such as the Mongol invasions across Eurasia or the Crusaders moving into the Near East (Zalloua et al. 2008; Zerjal et al. 2003). These and other examples will be discussed further in chapter six.

Autosomal Markers

In addition to the studies of uniparentally inherited markers, some autosomal markers have provided useful information to archaeologists. There are several mutations associated with specific environmental stressors, such as a collection of mutations resulting in a blood disorder known as thalassemia, which are particularly common in areas of high malarial endemism. While carriers of the thalassemia mutations tend to suffer from anemia, which is generally detrimental and would be selected against, this condition also confers a degree of resistance to malaria, which explains its high frequency in malarial zones. By studying the geographic distribution of these traits and, particularly, finding them at high frequencies in locations where there is no malaria, we can identify markers of population origin and spread. For example, several thalassemia markers are common among Polynesian peoples who live outside the range of malaria, indicating an ancestral relationship with populations in a malarial zone (Hill et al. 1985, 1987). This was some of the first biological evidence that clearly indicated Polynesian origins were in Near Oceania rather than the Americas. See Box 2.2 for a discussion of one method for identifying the sickle cell mutation using a method known as Restriction Fragment Length Polymorphism (RFLP) analysis.

Whole-Genome Analysis

The development of NGS technology and microarrays is allowing whole-genome surveys for variation to be undertaken. It should be noted that a "whole-genome analysis" is not necessarily sequencing the whole genome, but is more often surveying the genome for particular known polymorphisms. Because these types of studies are particularly subject to ascertainment bias, a number of consortia are developing regionally specific panels of markers for population studies. This massive and worldwide approach will undoubtedly increase the volume of information available for the reconstruction of population histories.

The Human Genome Organization, or HUGO, began when forty-two scientists from seventeen countries joined forces to coordinate human genome studies. The organization arose to share technologies, foster collaboration between scientists using different approaches to the human genome, and prevent the wastage of resources through the unnecessary duplication of efforts. Thus, the group aimed to streamline the work and to better understand the human genome. Further, HUGO is engaged in promoting public debate about human genome research. An example of the type of studies the group undertakes is a large pan-Asian SNP survey (Abdulla et al. 2009). These types of data are not, however, going to replace the work being conducted on mtDNA and Y chromosomes for a number of reasons. First, the SNPs under study may be subject to selection and recombination. They also do not capture recent population history as readily as do the uniparentally inherited markers. The use of SNP chips and assays of genome-wide STRs are good for understanding ancient relationships, while the more fine-grained studies (Friedlaender et al. 2008) indicate that linguistic and genetic exchange are particularly common and may obscure ancient patterns. As these approaches are combined in the future, we can perhaps shed further light on population histories.

Laboratory Methods

Obtaining DNA

DNA can be extracted from many different sources including blood, soft tissue (muscles, organs, tails, leaves, and fruit), hard tissue (wood, teeth, and bone), hair, skin scrapings, coprolites and fecal samples, eggshell,

Box 2.2: Restriction Fragment Length Polymorphisms—RFLPs

Restriction enzymes are proteins that cleave, or cut, DNA at specific base sequences (four, five, or six bases long) known as restriction sites. The specificity of cleavage allows for the low-resolution detection of sequence variation without resorting to the more time-consuming and costly business of sequencing. Mutations create or destroy restriction sites. Differences between individuals can be the result of point mutations in the restriction site, such that they no longer correspond to the cleavage sequence of a particular enzyme or larger scale changes in the sequence such as insertions, deletions, or rearrangements. Different sequences will, therefore, result in length differences among the collected fragments produced by each digestion, or the incubation of the DNA with a specific enzyme or combination of enzymes, which led to the name Restriction Fragment Length Polymorphisms (RFLPs). Individuals can be polymorphic in the length of their restriction fragments, thereby producing different patterns of bands on an electrophoretic gel.

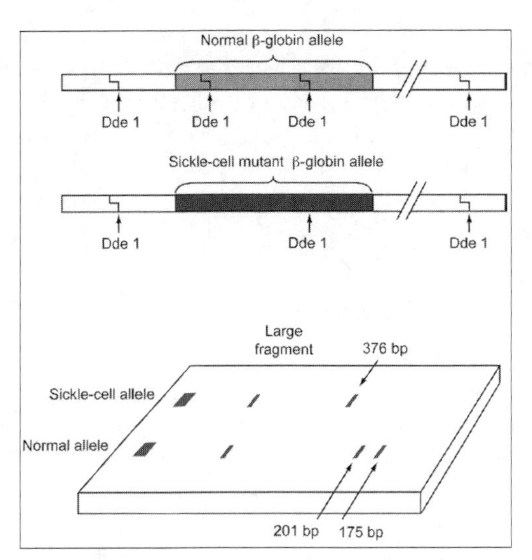

As shown in the figure, the mutation that causes sickle cell anemia (an A to T point mutation on the beta-globin gene) removes a restriction site that would normally be cut by the restriction enzyme Dde 1. The enzyme, Dde 1, cuts the DNA between the C and the T when it reads a DNA sequence CTNAG (where N can be any one of the four nucleotides). Normal hemoglobin sequences read CCT GAG GAG, which code for the amino acid sequence Pro, Glu, Glu. Whereas, the sickle cell mutation results in the sequence CCT GTG GAG, and thus codes for the amino acid sequence Pro, Val, Glu (see Figure 2.4 for the amino acid codes). The restriction enzyme Dde 1 does not cut the DNA at this site in individuals with the sickle cell mutation, and therefore, the RFLP pattern is different from individuals with the normal allele. When the digested DNA is run on an agarose gel, instead of having two fragments of DNA (one 175 bp and one 201 bp) that would occur in an individual with the normal allele, those with the sickle cell allele have a single fragment that is 376 bp in size.

and saliva. Recent developments have demonstrated that DNA can be recovered from environmental samples such as ice cores (Willerslev et al. 2007), sea water (Venter 2005; Williamson et al. 2008), and most importantly for archaeologists, soil (Willerslev et al. 2003). DNA extracted from such samples has significant potential for the reconstruction of ancient biotas. While it has been somewhat naively suggested that bone samples are no longer necessary to answer questions about past environments and extinct species (A. Cooper in Stokstad 2003) and that soil samples alone can provide the necessary genetic information, there are still issues regarding stratigraphic control and reproducibility. The possibility of contamination in and through the samples still makes this approach experimental and of limited use for most archaeological applications.

There are several methods used for the extraction of DNA, the most appropriate of which vary based on the type of tissue, the quality of the DNA, the preferences of the investigator, and the facilities available in the lab. All of these methods involve bursting the cells (if cell structure remains) and recovering total genomic DNA (nuclear and mtDNA). Rather than analyzing the whole genome now, researchers typically target a particular region of DNA, such as the hypervariable region of the mitochondrial genome, through the use of a technique called the Polymerase Chain Reaction.

Amplifying DNA—The Polymerase Chain Reaction

The Polymerase Chain Reaction is one of the most basic and widely applied techniques in molecular biology. It allows one to take a low concentration of DNA extracted from a biological sample and literally make billions of copies of the part of the genome of interest. This method is somewhat like a photocopier of DNA—rather than taking the whole book out of the library, you can simply photocopy the pages or chapter in which you are interested. The concept was developed in the mid-1980s by Kerry Mullis, a researcher at the biotech company Cetus in Emeryville, California (Mullis et al. 1986). According to one story published in *Scientific American* (Mullis 1990), the concept of PCR came to him when he was driving through the redwoods in Northern California (though in other sources he credits the development of the concept to an LSD trip; of course, the two stories may not be mutually exclusive). Regardless of where Mullis was, he was apparently thinking about how DNA replicates naturally in a cell and then asked himself, "why not recreate this natural process and harness it to artificially synthesize DNA in the laboratory?" He won a Nobel Prize in 1993 for this idea.

The PCR technique takes advantage of the structure of DNA in which the information content is encoded twice, once in each of the two strands. Artificial DNA synthesis, or PCR, works from the same principle. Double stranded DNA is heated to break the bonds of the rungs of the DNA ladder (this is called denaturation). Each single-stranded half of the DNA molecule is then used as the template for the construction of a new strand. The synthesis of a new strand of DNA in undertaken by an enzyme known as DNA polymerase (the most common polymerase is known as *Taq*), which cannot begin at any unspecified position in the genome. The synthesis must be "primed" or started by a small section of double-stranded DNA. Additionally, DNA synthesis can take place only in one direction along a strand. Therefore, two PCR primers (often referred to as a forward and a reverse primer) are based on known DNA sequences and designed to flank the region of interest or the "target sequence." Each of the two primers binds, or anneals, to the appropriate DNA sequence on the appropriate strand. This creates the necessary segment of double-stranded DNA that the DNA polymerase can start building on. The synthesis, or "extension," of a new longer piece of double-stranded DNA then proceeds from each of the two primers working in opposite directions toward the center of the piece of DNA being amplified (Figure 2.9).

All that is needed to make this process happen is an initial sample of whole-genomic DNA, two synthesized PCR primers, some extra nucleotides (As, Ts, Cs, and Gs) that will build the new strands, and some DNA polymerase, the enzyme that starts the chain-building process. All of this is put into a small plastic tube that is then placed into a PCR machine, also known as a thermal cycler, which is not much more than a computer controlled heating block. When the tube is heated to 94°C for a brief period of time (less than a minute), the DNA unzips or denatures. When it is cooled to around 50°C (again for a period of less than a minute), the primers attach to the appropriate spot in the genomic DNA—identifying the beginning and end of the region of interest. When the tube is heated again to about 72°C, the DNA polymerase takes the free nucleotides and extends the primers with the appropriate letters to rebuild, nucleotide-by-nucleotide, the double-stranded DNA. Every C binds to every G and a T to every A along the original single strand, creating two new double strands of that portion of the DNA. Thus, after one cycle of PCR, the amount of target DNA has doubled, and only that bit of the total DNA is in the tube. By repeating the heating and cooling cycle numerous times, a process that takes a couple of hours, enough of the target DNA sequence is produced to support analysis. The DNA polymerase enzyme will eventually exhaust itself after a number of heating cycles, but generally the PCR process generates enough DNA for most purposes after about thirty cycles. If there is only a tiny amount of target DNA, as there often is in studies of highly degraded aDNA, more PCR cycles may be necessary.

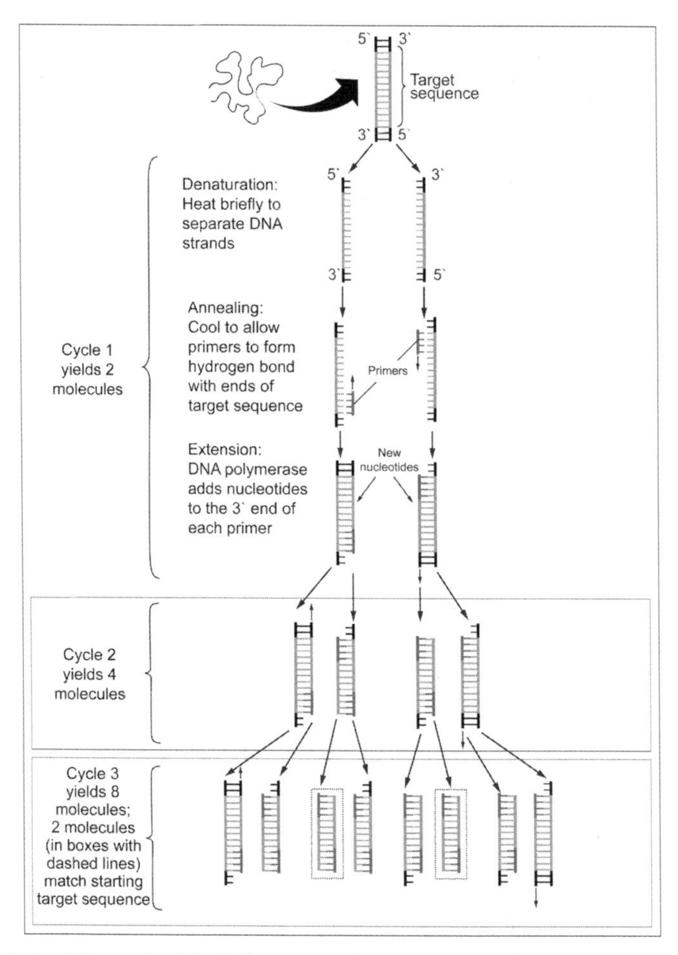

Figure 2.9 Schematic of the Polymerase Chain Reaction (after http://openwetware.org/wiki/Image:JCATutorial_PCRDiagram.jpg).

Once PCR products are generated, they are then processed through gel electrophoresis to confirm that the PCR was successful and the appropriate size fragment was amplified. After this is determined, the PCR product can be further processed to determine the DNA sequence. Today, this is most commonly performed with a DNA sequencing machine, which results in the production of the DNA sequence in electronic format, ready to compare with other DNA sequences (see Boxes 2.3 and 2.4 for a description of DNA sequencing techniques and gel electrophoresis for description of gel electrophoresis and DNA sequencing techniques).

Box 2.3: DNA Sequencing

Conventional, or Sanger, DNA sequencing works much like PCR. Each base is tagged with one of four fluorescent dyes with one color for each given base, A, C, T, or G. Each of the fluorescently labeled bases is also known as chain-terminating base because once they have bound to the extending DNA strand, they are chemically incapable of binding to another nucleotide. As such, they prevent any further synthesis of the new DNA strand, thus terminating the chain. During the extension phase of PCR when the A-dye-labeled terminator is incorporated into the growing chain, the chain is simultaneously terminated and labeled with the specifically matched dye. This continues until the entire product is amplified. The reaction products are then run through a capillary in a process similar to gel electrophoresis. As the individual dye-labeled fragments move down the capillary, they reach a point where their florescence is detected by a laser camera and the appropriate nucleotide is recorded—A, T, C, or G.

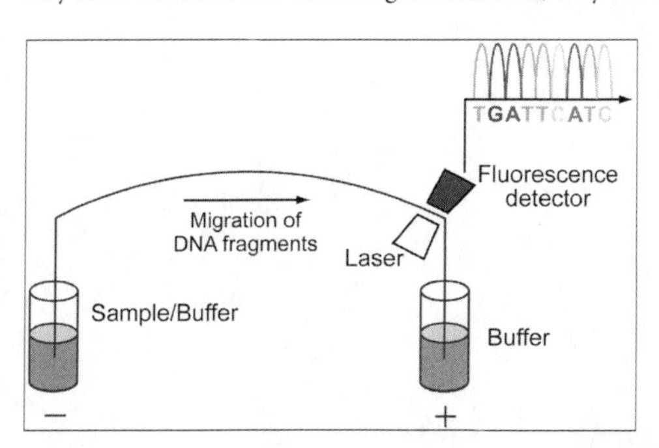

Analytical Methods

Phylogenetic Reconstruction

As discussed at the beginning of this chapter, DNA data can be used to address several important questions in archaeology. A large number of those questions relate to identifying evolutionary relationships between populations or organisms through time and across space. A phylogenetic tree, or a phylogeny, is a graphic representation of evolutionary relationships. Phylogenetic trees can be drawn using different types of data, including morphological and behavioral traits and, most importantly for the topic of this book, molecular data such as DNA sequences. Central to the concept of a tree is that of shared common ancestry. All phylogenies are hypotheses about the order of the evolutionary events—reconstructions of the past events that could possibly explain the variation present today.

So, how does an archaeologist get from a bunch of DNA sequence data to a phylogenetic tree? There is no simple answer to this question. For someone who is probably NOT going to actually analyze genetic

Box 2.4: Gel Electrophoresis

After samples have been amplified with PCR, an electrophoretic gel is "run" to determine the success (or otherwise) of the reaction. This is a method for checking the presence of an amplified fragment of DNA, and that it is of the expected size.

An **agarose gel** is used to size fractionate DNA fragments. An aliquot, or small subsample, of the PCR mixture is combined with loading dye (this allows the sample to be seen as it is "loaded" onto the gel and as it runs across the gel). It is then placed in a well of the gel, which is covered by a liquid buffer, and an electric current is applied to move the DNA across the gel. The negatively charged DNA is attracted toward the positively charged electrode and moves across the gel in that direction. The rate of migration through the gel is proportional to the size of the fragment, and DNA fragments of different sizes will produce bands at different distances from the wells. The technique can also be applied to proteins (and was, in fact, developed for proteins before being used for DNA), but their pattern of migration is more complicated than that of DNA because proteins fold into three-dimensional shapes, making their rate of migration dependent on properties other than length of the fragment. Size fractionation of DNA fragments is similar to using nested sieves for screening during archaeological excavations.

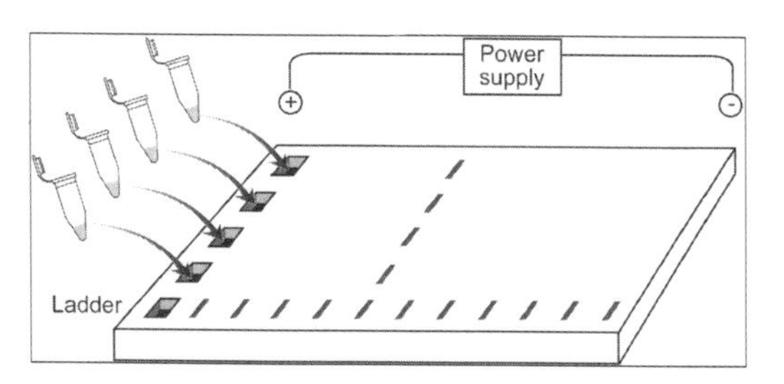

In order to actually visualize the DNA, some chemicals such as ethidium bromide or "SYBR Green" are added to the agarose. These chemicals bind to the DNA and fluoresce when exposed to UV or other appropriate light. This way, the DNA can be seen on the gel once the gel has been run and then placed on the light source.

DNA samples are run next to standardized DNA size markers to estimate the size of the fragment being run on the gel. These are called DNA ladders and come with a wide variety of size fragments to allow the determination of the size of very large DNA fragments that are many thousands of bases long or the tiny fragments that are typically recovered from archaeological samples.

data, it is not necessary to know the specifics. Archaeologists should, however, know that in order to build a tree, numerous decisions have to be made. Further, the person doing the analysis should be able to explain which decisions were made and why. When you read a genetics paper, much of this information will be contained in the methods section and sometimes not even in the paper itself, but in supplementary information. Here, we present a brief list of questions you should ask yourself when you read a DNA paper or that you should ask any DNA researcher with whom you collaborate.

How Much Sequence Data and What Data are Being Analyzed?

Important factors that must be considered in any DNA study are the length of the DNA fragment being studied, the number of sequences/samples being analyzed, and the kinds of markers (mtDNA, Y chromosome, and so on) being analyzed. As mentioned previously, it is now increasingly common, and some might say even necessary, when looking at mtDNA variation that complete mitochondrial genomes are studied. This is particularly true for studies of modern population variation. With the new NGS technologies available, we expect that even aDNA studies will be using these methods. Thus, larger and larger aDNA data sets will be published in the future.

What Methods are Being Used for Tree Construction, and What are the Evolutionary/Theoretical Implications of Those Methods?

We do not have all of the sequence of all the DNA of all the people living in the world today or in the past, so we must fill in the gaps and estimate our evolutionary history. In most cases, we actually have so few data that there are MANY possible trees that can be constructed from the data that are available. This is particularly true when reconstructing phylogenies of very closely related individuals, such as those within a single species. Determining the identity of the "best" tree is difficult. The most common algorithms used for the calculation of phylogenetic trees are known as **maximum parsimony**, **neighbor-joining**, and **maximum likelihood**. These can be undertaken using standard computer programs for phylogenetic reconstruction such as Phylogenetic Analysis Using Parsimony (PAUP; Swofford 1991), the PHYLogeny Inference Package (PHYLIP; Felsenstein 1993), and Molecular Evolutionary Genetics Analysis (MEGA; Kumar et al. 2008). For indepth discussions of these different methods, their strengths and weaknesses, and the various computer packages available for undertaking phylogenetic analyses, we suggest *Phylogenetic Trees Made Easy: A How-To Manual* (Hall 2011) or *Bioinformatics for Dummies* (Calverie and Notredame 2006). However, here we will present a basic description of these most-common methods of analysis.

Phylogenetic trees can be constructed with many types of data and can incorporate a range of evolutionary models. Data can be analyzed using two different approaches: 1) genetic distance methods; or 2) character state methods. Distance methods are based on similarities and differences calculated for each possible pair of sequences in a data set, thereby reducing the raw data to a series of pair-wise differences. These are the simplest and fastest trees to construct, but in themselves, they do not evoke particular evolutionary models. Examples of distance tree methods are neighbor-joining (Fitch and Margoliash 1967) and the unweighted pair group method with arithmetic means (UPGMA), also known as an average linkage method (Sokal and Michener 1958). The latter method, UPGMA, assumes a constant rate of evolution, while neighbor-joining allows branch lengths to reflect the amount of change along a lineage in which greater genetic distance results in a longer branch length.

Character states are discrete units of evolution such as nucleotide data, or the numbers of allele repeats in an STR data set. Character state methods analyze changes in character states in the different samples being studied. Building character state trees requires the identification of a specified evolutionary model, such as maximum parsimony, under which the tree is constructed. Biologists often use the concept of **parsimony** which assumes that the tree that requires the fewest number of changes or mutations, described as the shortest tree, is the most likely to be evolutionarily correct. Character state analyses also allow one to infer an ancestral state at any particular branching point on the tree.

More advanced methods involve an optimality criterion, and these are much more time consuming and computationally complex; although the use of these methods are becoming increasingly popular as computer power increases. One of the most common optimality methods is maximum likelihood analysis in which a calculation is made for determining the likelihood of generating the observed data, given a particular tree (Felsenstein 1981). This method works in the opposite direction from the distance methods. Distance methods work from the data to the tree. Maximum likelihood methods test particular trees and branch lengths in combination against the available data, searching for the tree most likely to have generated the observed data.

Bayesian methods are character state methods that allow the incorporation of prior knowledge into analyses of a data set and allow for the manipulation of variables to test specified demographic models against the genetic data. Bayesian methods employ a likelihood approach to determine a probability distribution for a set of all plausible trees or hypotheses for any given data set. The programs, Bayesian Evolutionary Analysis by Sampling Trees (BEAST, Drummond and Rambaut 2007) and MrBayes (Huelsenbeck and Ronquist 2001), are two popular and very useful software packages available for Bayesian analyses. These and other genetic population, model testing approaches are becoming increasing valuable for evaluating models of population migration and expansion.

Over the last few years, the developments in bioinformatic analyses have been so significant that it has now become difficult for even molecular biologists to perform all the necessary analyses on their data. It is becoming more common, and even necessary, to have a dedicated bioinformatician as part of the team involved in DNA sequence analysis. However, regardless of the sophistication of the analytical techniques, the quality of the data remains paramount. Therefore, all of the DNA data and archaeological information, such as radiocarbon data or archaeological context, if such information is incorporated in the model testing, need careful assessment by both the molecular and the archaeological teams.

Is the Tree a Rooted Tree or an Unrooted Tree?

The shared common ancestor of all individuals on a tree can be determined by comparing all of those sequences with one obtained from a closely related taxon, termed an **outgroup**. The use of the outgroup provides a "root" for the tree, identifying which sequences are most closely related to a more distant, known relative. For example, the human mtDNA tree produced by Vigilant et al. (1991) was rooted by comparing its sequences with those of our next closest living relative, the chimpanzee. Thus, the changes in the DNA sequences are given directionality through time.

Is the Tree Really a Tree, or is it a Network?

Phylogenetic trees force data into an assumed evolutionary relationship, but not all data may be tree-like due to recombination, hybridization, or horizontal gene transfer. Alternatively, the data that are being analyzed may be insufficient to reconstruct a true evolutionary relationship or phylogenetic tree. While ultimately we can say that all organisms are evolutionarily related, the data that we collect may be problematic and may not conform to a tree-like structure. Recognizing the inadequacies of the data we currently have for addressing the questions that we are asking, **network methods** are being used more frequently. This is particularly the case for genetic analyses of closely related taxa (such as analyses of variation within a population or a species) because the phylogenetic signal may be less clear due to continuous gene flow (Huson and Bryant 2006). Networks allow the noise and inconsistencies in the data to be faithfully represented, such that two equally well supported trees can both be displayed in a single network. Then, an arbitrary choice need not be made about which is more likely to be correct (Figure 2.10). Networks summarize all of the most likely possible reconstructions and can be calculated from either distance or character state data.

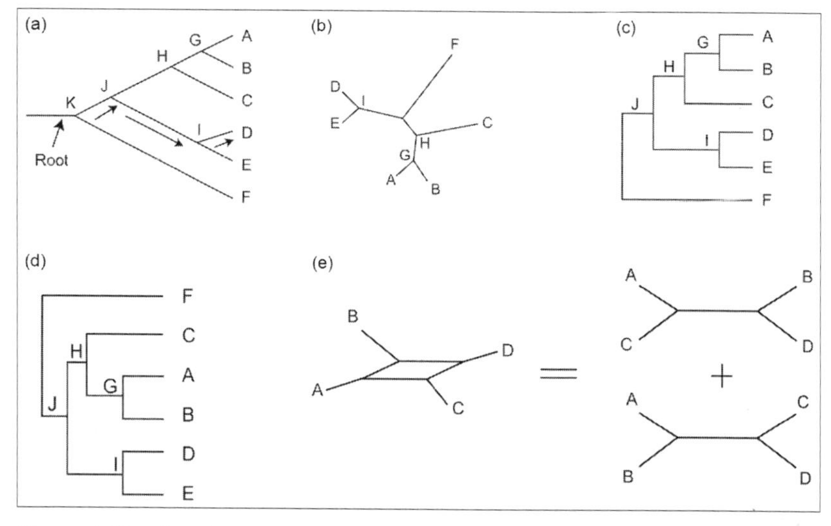

Figure 2.10 A series of representations of phylogenetic relationships. (a) The first is a rooted phylogenetic tree displaying directionality, or time. The same relationships are shown in (b) but the tree is not rooted and therefore does not indicate the MRCA. The phylogenetic trees in (c) and (d) display exactly the same relationships between taxa. The information content of each tree is identical despite their different appearances. Representation (e) shows how two different but plausible sets of evolutionary relationships can be displayed in a single reticulated network (after Vandamme, 2009, Figures 1.4, 1.5, and 1.8).

Confidence and Statistical Support for Trees

There are almost always many equally parsimonious trees for any given data set and these can be summarized in a **consensus tree** in which all branches have support at a determined level. In a 50% consensus tree, all the branches supported by 50% or more of the trees are shown, and all of those branches that do not meet the 50% threshold are collapsed. Another way of assessing the reliability of a particular tree or network is to test whether the reconstructions consistently converge on the same patterns. **Bootstrapping,** for example, generates new data by resampling and replacing the available data set (Efron and Gong 1983). Bootstrap values represent the percentage of times that a particular split on a tree is seen in a defined number of replicate resamplings. These values will often be recorded on or above particular branches of a tree. Generally, it is recognized that lower bootstrap values suggest less statistical support for particular branching relationships. The actual cutoff point for reliability is arbitrary, but bootstrap values of 95% or more are often considered indicators of strong support (Nei and Kumar 2000). It should also be noted, however, that high bootstrap values do not necessarily indicate the level of accuracy of the reconstruction, merely the tendency of the available data to show the pattern. The assumption employed in bootstrap analyses is that if the data are genuinely indicating a particular tree, then a resampled subset of those data should identify the same relationships. The data themselves can be an inaccurate reflection of prehistory but still produce high bootstrap values.

Interpreting Trees

Most of the phylogenetic trees that the reader may have seen are probably gene trees. In other words, the trees were built with data derived from a single region of the genome. Just as trees constructed from mtDNA and Y chromosome data are not identical, phylogenetic trees built using data from several different gene regions may result in different phylogenies. As with any other statistical study, the more data that are available, the greater confidence one can have about the inferences being drawn. Additionally, patterns that are consistent across data sets are more likely to reflect reality. Nevertheless, error can be introduced into phylogenetic reconstruction by numerous factors, including a complex evolutionary history, an inaccurate model of sequence evolution, and probably most commonly when dealing with archaeologically derived DNA sequences, insufficient data (Ives et al. 2007).

As much as we might hope to the contrary, a reconstructed phylogenetic tree does not necessarily lead straight to a specific, historically meaningful interpretation. Many different interpretations may be consistent with a given tree topology, or structure, and these possible interpretations must be evaluated within an appropriate context. Apparent conflicts with independent data, such as those from archaeological, linguistic, or ethnohistoric studies, must be given substantive consideration. In this regard, however, it must be noted that a molecular estimate for the date of a last common ancestor will not necessarily align with an archaeologically visible event. Genetic estimates for the dates of animal domestication, for example, are always significantly earlier than morphological or behavioral indicators of domestication or the archaeological evidence of the presence of domesticated animals because they will only indicate when wild and domestic populations last shared a common ancestor (Achilli et al. 2012; Vilà et al. 1997). In fact, this date MUST be earlier than the actual appearance of morphological indicators of domestication. In addition, domesticated plants and animals are rarely completely genetically isolated from the wild populations from which they were derived, unless they are moved to a region where they are not native. This possible wild/domestic interbreeding will have an influence on the error rates for calculations of the timing of domestication, causing them to be quite large. Thus, molecular estimates may not be useful for discriminating between archaeologically derived hypotheses about dates for particular domestication or dispersal events, particularly those that occurred in the relatively recent past.

One final point that we must always remember when looking at genetic data and inferring historical events or evolutionary relationships is that we are, for the most part, looking at gene trees. Gene trees are constructed using only one particular fragment of DNA. Different parts of the genome may have different rates of mutation and different inheritance patterns, which would, therefore, result in different trees. Similarly, within a population, different individuals will have different histories. Populations are made up of multiple individuals each with complicated histories and recombined genomes reflecting different aspects of their ancestry. The multiple stories gleaned from multiple markers are *all* real aspects of the ancestry of that person or population. Thus, the challenge is to reconcile the different patterns seen in the various genetic markers with the information available from linguistic and cultural heritage, which like different parts of the genome, can have different modes of transmission across generations (Pamilo and Nei 1988).

Genetic Dating

As previously discussed, rooting a tree results in the construction of a relative chronology of genetic change; but it is critical to keep in mind that an appropriate outgroup needs to be identified to root the tree. However, when trying to reconstruct population histories and past events, prehistorians are particularly interested in identifying an absolute chronology. We want to know when particular populations split or shared a common ancestor, so we can compare the genetic evidence with the archaeological or palaeontological data. In order to construct an absolute chronology, we can apply a concept known as a molecular clock (Bromham and Penny 2003; Zukerkandl and Pauling 1965).

A molecular clock requires calibration from independent information, meaning the accuracy of the clock depends on—among other things—the quality of the independent information. The most common calibration point that is used in reconstructing the history of the human species is the human-chimpanzee split of 5–7 million years ago (Hasegawa et al. 1985; Hasegawa et al. 1993). Calibration points within modern human prehistory have been used based on relatively well-accepted colonization dates of places such as Sahul (Australia and New Guinea) at 40–60 KYA (Hudjashov et al. 2007; O'Connell and Allen 2004). Clearly, this can result in a somewhat circular argument, with archaeological and genetic data being used to calibrate each other. Because molecular clocks are only as good as their calibration data, multiple calibration points should be used whenever possible.

There are a number of assumptions required to develop a molecular clock. These include a constancy of mutation rate, demographic factors such as population size, generation time, migration, and an absence of selective pressures on the genetic marker. Mutation rates can be reported in a number of different ways including the number of base substitutions per site per million years, or per site per generation. This often makes it difficult to reconcile results from numerous studies. A study of complete mitochondrial genomes in humans from around the world, assuming a generation time of twenty years and a human-chimpanzee divergence of five million years ago, calculated a mutation rate of 1.7×10^{-8} per site per year, which equates to 0.017 substitutions per site per million years (s/s/myr; Ingman and Gyllensten 2003). This calculation was for the mitochondrial genome, excluding the control region, which can be problematic for estimates of rate due to the high number of **parallel substitutions**. Different researchers use different estimates of generation time, usually ranging between twenty and thirty years, but note that their choices have an impact on the calculated divergence time. A study of the mutation rate of the control region alone was undertaken using data collected from 272 individuals of known Icelandic pedigree. This study found a mutation rate of 0.32 s/s/myr (Sigurðardóttir et al. 2000). Given that the actual phylogeny of the subjects was known, the researchers were able to get around the issue of parallel substitutions that plague standard population studies by examining mutation rates in the control region.

In human populations, clock-like behavior is most likely to be disturbed by common events such as population expansions, migrations, population crashes, and societal changes regarding the age at which the first offspring are born. One way of detecting population expansion is through the use of **mismatch distributions** in which pairwise differences between DNA sequences collected from within a population are calculated for all members of the sample (Figure 2.11). By comparing sequences and counting the number of differences between them, we can extract demographic information about a population (Schneider and Excoffier 1999). The frequencies of the pairwise differences are plotted on a histogram, and information about the population's history is gleaned from the mean value of differences between sequences and the shape of the distribution. The mean value of differences provides an indication of the overall diversity in the population moving toward greater values of difference as the population becomes older. The shape of the distribution is influenced by the stability of the population size; a smooth

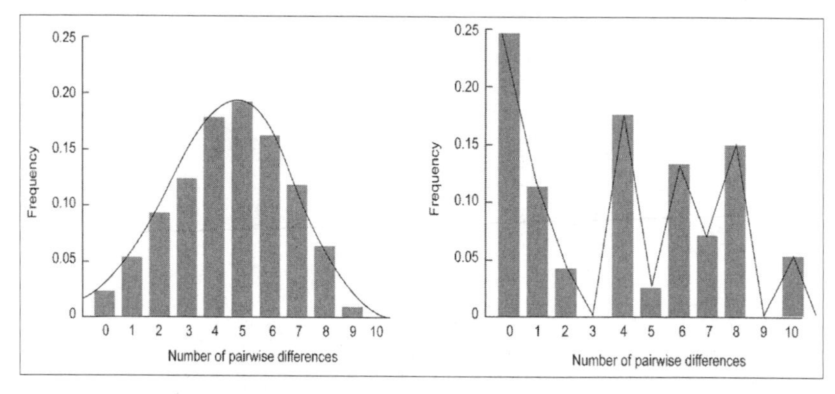

Figure 2.11 Mismatch distributions. Numbers of pairwise differences between sequences are plotted on histograms. The mean pairwise difference increases as the population ages, and the shape of the distribution changes in response to the stability of the population size. A smooth distribution reflects an expanding population, while a ragged distribution reflects a population size that has been stable (after Jobling et al. 2004:159).

bell-shaped curve indicates population expansion, and a ragged, multi-modal distribution indicates a long-term stable population size.

Migration is another demographic factor that will potentially affect the assessment of clock-like mutation rates. While migration has increased dramatically with twentieth -century transportation systems, people have always moved. Populations were replaced, and populations became extinct. The variation that we see on the landscape today does not necessarily reflect the occupants of a particular location in the past. However, now that it is possible to access DNA from archaeological material, we can sample genetic variation in a particular time and place to address the specifics of regional culture histories and larger patterns in prehistory.

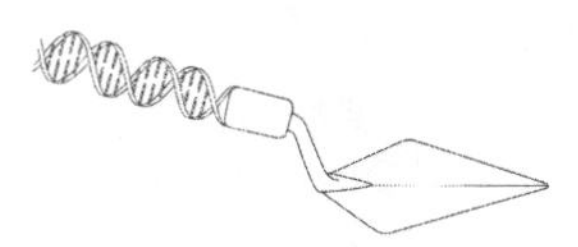

Chapter Three

Ancient DNA

This is a book written for archaeologists. We do not expect that our readers will be undertaking aDNA analysis themselves, so it might be asked why archaeologists should care about the laboratory environment or the procedures that are used to control and assess issues of paramount concern, such as contamination. The primary reason is that archaeologists are the first step in almost all anthropological aDNA studies. What happens in the field makes all the difference in the outcome of aDNA research. Furthermore, given the time and money necessary for the excavation of samples, it is worthwhile to anticipate aDNA studies before going into the field so that project personnel can be prepared to collect samples and information that might be critical in any future analyses. This is particularly important when working in an isolated location without a nearby university, grocery store, or other facilities where necessary equipment such as tubes or bleach can be obtained. There is also significant value in collaborating with potential aDNA specialists as early in the process as possible. Most importantly, archaeologists should discuss with them the questions that the project aims to address with genetic data. This will ensure that researchers come to understand the kinds of data that can be expected to emerge and what those data will, and (crucially) will not be able to say. In the interest of establishing productive collaborations, it is also worth discussing how much time it is likely to take to generate the data and if it is likely that no meaningful data will be recovered. It is our experience that it is easy for archaeologists to vastly underestimate the laboratory time necessary to generate useful data, and they may thus find the process unexpectedly frustrating. This chapter is intended to be practical, and it is particularly aimed at those who want to contribute to aDNA studies or use aDNA data to address their own archaeological questions.

DNA for Archaeologists by Elizabeth Matisoo-Smith and K. Ann Horsburgh, 59–79.

As briefly mentioned in chapter one, aDNA refers to DNA from any sample that is degraded or was not recently collected from a living individual and preserved. As soon as an organism dies, its DNA begins to decompose. By the time that billions of microorganisms have broken down the tissue and have left only the skeleton behind, just a tiny fraction of the original DNA remains, sequestered away in the bones and/or teeth. Over the years between burial and recovery by archaeologists, that DNA breaks down further—particularly if exposed to fluctuations in temperature or humidity. Consequently, aDNA research suffers from two major problems: 1) the gross concentration of DNA in an ancient sample is far less than that in fresh tissue; and 2) the fragments into which the surviving DNA has degraded are short (often under 100 bp long), making them difficult to analyze. In addition, once an archaeologist discovers skeletal remains or other samples from which they might want to obtain DNA, a new problem for the aDNA researcher begins—contamination of the sample by modern DNA.

The most important topic that should always be discussed when reading or writing about aDNA (particularly ancient human DNA) is the issue of contamination. Contamination refers to any DNA that is derived from sources other than the specimen being analyzed. DNA is present in every cell—in saliva, in dandruff, in other bodily detritus—and cells are constantly being sloughed from the body. From these modern sources, DNA is in excellent condition and will inevitably be transferred to anything we touch. To make matters worse, we are not only leaving *our* DNA everywhere, but we are also shedding bacterial cells and *their* DNA. That DNA is terrific for a forensic scientist who is interested in analyzing the DNA left on a cigarette butt or a beer bottle, but it is a nightmare for an aDNA researcher. Once excavated, archaeological samples are handled by many people in the field and then sent back to a museum or university where they are handled by more people as they are catalogued and studied. By the time someone decides to analyze them for an aDNA study, it is usually impossible to know how many different people have contributed their own DNA to the sample. Then, when the sample is taken to the lab for DNA extraction and analysis, PCR—the method used to target and amplify the DNA (see discussion in chapter two)—preferentially copies good-quality DNA over the degraded and low-quantity aDNA in which we are interested. In many archaeological remains, the DNA will be so diminished that a DNA extraction might contain only a few intact molecules, making it particularly easy for modern, contaminating DNA to out-compete the real target during PCR amplification.

In addition to modern DNA being sloughed from all those who handled a sample, amplified DNA from any PCR experiment escapes as an aerosol from laboratory tubes into the laboratory environment every time a researcher opens a tube. This DNA is not only transferred to the equipment and the gloves of the researcher, but also to the floor of the laboratory. From there, it gets on the shoes of the researcher and anyone else walking through the lab and is tracked throughout the building. Thus, the molecular anthropologist is trying to recover the few remaining copies of endogenous or target DNA from the archaeological specimen, when the sample can also potentially contain billions of molecules of contaminating DNA.

In light of the preceding discussion, one might think that the ideal situation is that human hands never touch any recovered material that might be analyzed for DNA. Until recently, most aDNA researchers would have agreed with this view. However, a recent study was conducted in which three teeth were removed from the intact, alveolar processes of ten Danish, Viking-aged skeletons (Melchior et al. 2008). Specialists in aDNA retrieved two of the three teeth of each individual from in situ contexts, and the third tooth was collected by physical anthropologists after osteological analysis. All of the researchers kept in mind the fact that the sample would be used for aDNA studies. For each individual sample recovered in the field, molecular anthropologists wore full bodysuits, hairnets, facemasks, shoe covers, and gloves while extracting the two in situ teeth from the deceased. In each case, the third tooth was then removed by a researcher wearing gloves and a facemask. Further, all individuals working at the archaeological site, in the museum, or in the molecular laboratory had their DNA sampled and analyzed. The aDNA from the teeth was extracted and processed by different researchers, at times separated by several days. The PCR products were amplified and cloned (this process will be described and explained later). At least eight clones were sequenced for each PCR amplification. Identical results, indicating only a single sequence variant, and therefore, a single individual as the source, were obtained for each pair of "clean" teeth extracted by molecular anthropologists (no contamination was detected). In each case, the third "handled" tooth for each individual produced numerous sequences, indicating the presence of DNA from multiple individuals. The truly scary result is that none of the DNA sequences obtained from the tooth matched any of the researchers involved in the project, nor were any of those sequences found in DNA extracted from the soil samples collected at the time of excavation.

The authors of the paper could not identify the sources of the contaminating DNA. It is possible that it was introduced from equipment used during the excavation and cleaning of the bones. Although contrary to common practice, the project personnel did not wash the bones. Washing is a known of source of contaminating DNA (Gilbert et al. 2006).

They were only brushed clean. It is possible that archaeologists were carrying around degraded DNA from other excavations or other archaeologists on their tools. This hypothesis deserves further attention, but in the meantime, cleaning archaeological tools with bleach (though you should be aware that bleach is corrosive and therefore may damage your trowel) or using disposable plastic or wooden tools for the recovery of samples for DNA analysis may be prudent.

Given the above discussion, it is not surprising that molecular anthropologists working with aDNA can be paranoid about contamination and can often be identified by the smell of bleach, one of the favorite methods for destroying contaminating DNA. Readers should be aware, however, that reliable ancient human DNA can be recovered from archaeological remains, and aDNA researchers have a number of methods they can employ to circumvent the problems of contamination of material that was previously excavated under less than ideal conditions. Yet, this example highlights the value of early planning and collaboration between archaeologists and molecular anthropologists. What follows is a discussion of a number of other important issues that archaeologists should be aware of.

Preservation and Ancient DNA

Visual inspection of a bone cannot reveal whether it has retrievable aDNA. While there are no systematic data to support the contention, it seems DNA is unlikely to be preserved in particularly fragile and friable bone. However, bones with particularly beautiful visual preservation or even samples that may still have skin attached do not necessarily retain DNA. Various characteristics of the environmental context in which the bone was found and the age of the material influence the level of degradation. Before seeking DNA, researchers often apply relatively cheap and quick biochemical techniques to determine the level of organic preservation in archaeological remains, such as amino acid racemization (but see Collins et al. 2009; Poinar et al. 1996), histology (Colson et al. 1997; Jans et al. 2004), mineral alteration (Götherström et al. 2002), mass spectrometry (Poinar and Stankiewicz 1999), collagen alteration (Koon et al. 2010), and osteocalcin analysis (Smith et al. 2005). Collagen is often extracted from archaeological bones to obtain an accelerated mass spectrometry (AMS) date, and if no collagen is obtainable, it is unlikely that there will be any DNA preservation. Thus, if it cannot be dated, it cannot be sequenced.

The integrity of DNA in living organisms is maintained by repair enzymes. Once an organism dies, those enzymes cease to function, and the DNA begins to disintegrate. Bacteria, fungi, and insects can accelerate the process. While aDNA is degraded by definition, some environments are better for DNA preservation than others. The ideal condition is permafrost. Essentially, the samples remain in a freezer. Granted it is one with a lot of microbial life, but nonetheless, it is a freezer. Cold, dry, cave sites also provide good conditions for aDNA preservation. Fluctuations in temperature, humidity, and pH are all problematic for DNA. While we don't fully understand how the pH of the soil affects DNA preservation, highly acidic soils destroy bones, making DNA preservation unlikely. Preservation is worst where there are high heat and humidity and in sites where water percolation occurs. There is generally a correlation, therefore, between site type and DNA retention (Dobberstein et al. 2008; Kaiser et al. 2008; Robins et al. 2001). In cold, dry conditions, DNA has been obtained from bones that are tens of thousands of years old (Gilbert et al. 2008; Hofreiter et al. 2007; Knapp et al. 2009; Krause and Briggs et al. 2010; Krause and Fu et al. 2010). However, in the hot and wet environments of the tropics, we may be limited to working with samples that are no older than a few hundred years (Robins et al. 2001).

As a general rule of thumb, aDNA researchers may conduct initial work on relatively less important samples to ascertain the likelihood that DNA remains in more precious or significant specimens, such as human remains. If DNA cannot be obtained from animal bones from a site, then researchers may not attempt to extract DNA from human remains, as this would mean destroying valuable specimens with little hope of DNA recovery. Some studies have suggested, though, that DNA can be preferentially preserved in bones of certain species, particularly avian bones, perhaps because of the nature of the lamellar structure (Haynes et al. 2002). An additional factor that must be taken into account is that DNA preservation can vary across a single archaeological site. Thus, there are no hard and fast rules regarding the likelihood of DNA preservation in a particular site or in a particular sample—sometimes you just have to try and see.

Types of Tissue and DNA Preservation

Different tissues are known to preserve DNA differently. Structures that protect the biological remains from water and microbial attack are desirable, so teeth are particularly good for aDNA work because the hard enamel of the tooth protects the DNA-containing pulp. However, if the enamel is damaged because of caries or postdepositional decay or damage, then the level of protection is reduced. In general, for aDNA retrieval, cortical bone is a more likely source than spongy bone. In addition, researchers may want to consider bones for aDNA analyses that will not be particularly valuable for other studies given the destructive nature of DNA extraction.

Sometimes molecules derived from a burial context, such as humic and fulvic acids, which come from decaying vegetation, will be coextracted with the DNA of the specimens (Baar et al. 2011). These can interfere with the action of the DNA polymerase that synthesizes new DNA during PCR amplification, thereby inhibiting the reaction. There are several additives that can be used during extraction or PCR that can help overcome this inhibition. It is a good idea to provide aDNA researchers with any information that you might have regarding the burial context in order for them to determine relative likelihoods that a lack of PCR amplification may be due to inhibition or poor preservation.

The Ancient DNA Laboratory

Once a specimen has been identified for aDNA analysis, utmost care must be taken to ensure that the bone goes nowhere near amplified DNA products. In practice, this means it must never be taken into a building that houses a PCR machine, which includes most departments of biology, genetics, microbiology, botany, and medical schools. The aDNA laboratory, therefore, should ideally also be located in a building where no amplified DNA will be found, or it should be in a part of the building that is completely isolated from PCR work. Ancient DNA labs and everything in them must be protected from contaminating DNA, which means that all equipment and chemicals arrive at the aDNA lab without passing through a modern lab. Any researchers entering an aDNA lab should be careful to make sure that they are not carrying any amplified PCR products into that lab. This means that they cannot go from a building with a modern DNA lab into an aDNA lab later in the day. Unidirectional flow must be maintained such that researchers planning on doing any aDNA work must consider the likelihood of transferring modern and amplified DNA into the aDNA lab.

Steps that different labs take to avoid introducing DNA into the aDNA lab vary, but all address the risks associated with DNA adhering to shoes, clothing, and hair. In our case, when we enter the lab, we remove all street clothes and shoes, which are left at the door. After donning scrubs to enter an outer room in the lab, dedicated lab shoes, coveralls, facemasks, hairnets, and two pairs of gloves are put on before we enter the lab proper. The aDNA lab itself maintains positive pressure, overhead UV lighting, and bleachable surfaces. Bleach is the aDNA researcher's best friend—it destroys DNA and can be liberally applied to all lab surfaces, removing a potential source of contamination (Dissing et al. 2008; Gilbert et al. 2006; Kemp and Smith 2005; Malmström et al. 2007). Ultraviolet light (UVC rays) also breaks down DNA so it cannot then be amplified (Gilbert et al. 2006). Thus, having overhead

UVC lighting that is turned on to decontaminate the lab once researchers leave the room is ideal. Unfortunately, UV light also damages plastics, so care must be taken to choose equipment that is UV-resistant, and any susceptible surfaces (such as tubing or plastic components to machinery such as centrifuges etc.) should be covered in foil for protection. All chemicals and containers used for aDNA should be disposable and guaranteed to be free of DNA by the manufacturer.

Our advice to archaeologists is this. When considering giving a sample to someone for aDNA analysis, ask them about their lab. Similarly, when reading an aDNA paper, there should be some discussion about the laboratory location and the protocols used to address the issue of contamination. If there isn't a discussion of protocols, or if the work is done in a lab that is not a specifically designated an aDNA lab, do not trust the results (see discussion in Knapp et al. 2012).

Processing Samples for Ancient DNA

If the specimen for analysis has not already been photographed, then a photographic record should be made because aDNA analyses are destructive. Archaeologists should also keep in mind the destructive nature of the analyses when choosing specimens for aDNA studies. Methods have been developed to preserve the external appearance of particularly important bones, such as those on display in museums, but they are expensive, time consuming, and appropriate for only some elements. Therefore, they cannot be considered a routine protocol. The use of dental drills has been frequently applied to this end. However, recent studies have shown that the heat generated by the drilling itself may cause significant damage to any remaining DNA (Adler et al. 2011). After photographing the specimen, any possible surface contaminants need to be removed from the bone or tooth. Bleach is particularly good for removing surface contamination from teeth because the enamel effectively protects the DNA inside (Kemp and Smith 2005). Bone surfaces can initially be wiped with bleach. The surface of the bone can then be physically removed with sterile sandpaper, or it can be scraped away with a scalpel. Both teeth and bones can be subjected to UV irradiation to damage any contaminating surface DNA. A portion of the specimen is then pulverized in a sterilized mortar and pestle, coffee grinder, or Cryomill (a liquid-nitrogen-cooled, magnet-driven, smashing device). When undertaking DNA analysis of an ancient specimen that could provide a particularly significant result, enough of the sample should be maintained for an independent replication and possibly a direct radiocarbon date. In addition, researchers may want to consider if other biochemical analyses may be undertaken so that subsamples can be taken and set aside for those analyses. Other analyses commonly include the extraction of stable isotopes for dietary reconstruction. This subsampling should, however, take place in the aDNA lab whenever possible given the requirements for unidirectional flow of materials.

Once the specimen has been reduced to powder, the extraction process takes place. A tube, in which no sample is placed and which is known as the negative control, is processed alongside the aDNA specimens to detect the possibility of introduced DNA from reagents (chemicals used in the extraction process). The negative control should be stored for all future analyses of the relevant sample.

The first step in the aDNA extraction involves digestion of the bone powder with enzymes that will break down the cells, releasing the total genomic DNA into a solution. A portion of this digest is then processed to actually extract the DNA and wash away all other cell components. Often, depending on the amount of bone that was ground up, a single digest will produce enough volume for several extraction procedures, and the digested bone powder can be stored in a freezer for future use.

Once the DNA has been extracted, it is in solution and available for future analysis. Typically, a particular region of the DNA of interest, such as the hypervariable region of the mtDNA, is amplified using PCR. In addition to submitting the negative extraction control to PCR, an additional negative control is set up to test for contamination of the PCR reagents. Multiple PCRs can be performed with the results of a single DNA extraction, but the extract volume is finite, limiting the total number of reactions possible.

In addition to amplifying the DNA from the actual sample of interest, positive controls, involving the amplification of a sample that is already known to contain amplifiable DNA, are often included in a PCR study to confirm that the all reagents and the PCR process itself did work. A non-result could mean that there is no DNA in the sample or it could mean that something went wrong with the PCR process. In an aDNA study, if a positive control is necessary, it should consist only of DNA extracted from an ancient sample that has already been sequenced and that can be readily identified as a possible contamination source if it is found in any of the test specimens or in the negative controls. Specifically, the positive control for an aDNA study should not be DNA extracted from a modern sample, as extracted modern DNA should NEVER be in an aDNA lab due to the possibility of cross-contamination. You can, therefore, see how the costs of aDNA research begin to accumulate. For each extraction undertaken, at least two additional samples for PCR must be processed. These include the negative extraction control, the negative PCR control, and, in some circumstances, a positive PCR control. Added to this is the fact that for each result, it is necessary to replicate that result to confirm that contaminating DNA was not amplified and sequenced.

All of the set up for the PCR amplification of ancient samples should be conducted in the aDNA lab. However, once set up, the sealed tubes are removed from the aDNA lab for actual amplification. In other words, the PCR machine being used for amplification should not be located within (or near) the aDNA lab itself. All post-PCR work including the preparation for DNA sequencing should be conducted in an external lab. Once the PCR process is finished, the product (the amplified DNA) is run on an agarose gel for visualization, which will confirm that the PCR itself worked and that the amplified DNA fragment is of the expected size and quantity. The degraded nature of aDNA means that only small fragments should be routinely expected.

In general, aDNA extracts permit only the amplification of fragments shorter than a few hundred base pairs. Recovery of longer fragments generally indicates contamination with high-quality, modern DNA (Hofreiter et al. 2001). Therefore, in order to check for possible contamination, researchers will often try another PCR amplification of the sample using PCR primers that will amplify large (for example, greater than 500 bp) products. If there is no contamination with modern DNA, then this amplification should not work. Generally, in aDNA research, the onus is on the researcher to prove that the results obtained for a sample are not the result of contamination. There are numerous steps and standard protocols required at various stages from extraction to sequence analysis to satisfy these requirements (which are discussed later in this chapter), and these should be mentioned in any reliable aDNA research paper.

Obtaining the actual DNA sequence data from the amplified PCR product can be done using several methods. If enough PCR product has been generated, then it can be directly sequenced, which involves simply cleaning up the PCR product (removing salts and other reagents) and sending that entire product to a DNA sequencing facility or sequencing it with any number of available machines in-house. Alternatively, the PCR products can be cloned into bacteria during which a single PCR-produced molecule is incorporated into the genome of a bacterium that will replicate itself naturally, thus copying the inserted PCR product along with the rest of its own genome. The amplified portion of the genome is then removed from the bacterial genome and sequenced. The important distinction between direct sequencing of a PCR product and the DNA copied by the bacteria is that direct sequencing reads the dominant sequence of all amplified DNA in the PCR reaction regardless of the possibility that multiple DNA sources, and therefore, multiple sequences may have been present in the original reaction. This means that the authentic, but rarer, aDNA sequence from the specimen might be missed, and the DNA sequence of a contaminant or a damaged fragment with misincorporated nucleotides will appear to be a new sequence variant. In contrast, when PCR products are cloned, generally several clones are sequenced, providing a window into the possible different DNA sequences in the original PCR amplification. Additionally, sequencing multiple clones allows patterns of DNA damage to be assessed. The most common form of DNA damage leads to changes from C to T and G to A (Bandelt 2005; Binladen et al. 2006; Bower et al. 2005; Briggs et al. 2007; Brotherton et al. 2007; Gilbert et al. 2005; Gilbert and Hansen et al. 2003; Gilbert and Willerselv et al. 2003; Green et al. 2009; Ho et al. 2007; Hofreiter and Jaenicke et al. 2001). Because aDNA PCRs begin from very few starting molecules, any DNA damage resulting in such a misincorporation among one of those few starting molecules will populate a PCR product pool with the descendants of that original damaged molecule and appear to be a mutation. By amplifying the extracted DNA multiple times and sequencing individual clones, researchers can see the pattern of base change, discern the real sequence, and eliminate those resulting from DNA damage. Unless hundreds of clones are being sequenced though, rare variants are difficult to identify and may be the real aDNA sequence.

A new technology called **pyrosequencing** has been developed in the last few years and is now increasingly used in aDNA research (Knapp and Hofreiter 2010; Millar et al. 2008). Pyrosequencing is also referred to as NGS, sequencing by synthesis (SBS), high-throughput sequencing, or massively parallel sequencing. Pyrosequencing eliminates the need for cloning because it sequences virtually all of the individual DNA molecules from the PCR. It is, in effect, in vitro clonal sequencing. It is as if thousands of mini PCR tubes, in which only a single molecule of DNA is being amplified, are being processed at once on a single pyrosequencing run. While financial constraints have generally limited the numbers of bacterial clones that labs have been able to sequence for a single specimen, pyrosequencing permits the cost-effective sequencing of hundreds, thousands, or even millions of PCR molecules per specimen. This fact vastly enhances the likelihood of recovering all DNA sequences in a sample, from which it may then be possible, often with the help of high-speed computation, to sort out those that are likely to be contaminating sequences. While still not yet widely applied in aDNA studies, largely because of their high start-up costs, pyrosequencing technologies are becoming increasingly available and affordable. Their application will no doubt continue to spread in the aDNA field.

Until the development of these new pyrosequencing methods, most aDNA studies exclusively focused on mtDNA. While samples that are well-preserved or not very old have yielded nDNA using traditional PCR methods, as is necessary for determining the sex of ancient remains, the multiple-copy number of mitochondrial genomes in each cell has allowed mtDNA to be recovered from ancient samples more frequently than nDNA. One major advantage to pyrosequencing is that, theoretically, it can recover all of the extracted DNA from a sample. Unlike PCR, it is not necessary to use primers to target the DNA of interest—all of the DNA that was extracted from the sample will potentially be amplified. When working with highly degraded samples, there is often insufficient DNA in the ancient extract to provide enough starting molecules for direct pyrosequencing. When this is the case, the sample can be increased through a step involving a pre-pyrosequencing amplification. One of the major problems with aDNA extracts, however, is that there is often a lot of environmental DNA (from bacteria and other organisms that were present in the burial environment) extracted with the DNA of interest. In the first major pyrosequencing study of permafrost-preserved, mammoth tissue, Poinar et al. (2006) found that less that 50% of the extracted DNA was actually from the mammoth. Another pretreatment for degraded or "dirty" extracts that may contain a large amount of environmental contaminants involves cleaning up the sample through a process known as **hybridization** in which the target DNA is fished out of the overall extract by causing it bind to a probe made of single stranded DNA from the same or a closely related species (Maricic et al. 2010). The probe contains specific protein labels attached to the DNA that allow the probe and its bound DNA to be held while the rest of the DNA is washed away.

There is yet another advantage to the new pyrosequencing techniques for aDNA studies. Conventional DNA sequencing (also known as Sanger sequencing, see Box 2.3) is limited to analysis of DNA fragments of 80 bp or larger. If the extracted DNA contains fragments smaller than 80 bp, which studies have shown is commonly the case for aDNA (Briggs et al. 2007; Briggs et al. 2009), those fragments are inaccessible to traditional approaches. Pyrosequencing, on the other hand, actually requires small fragments of DNA. Indeed, when working with modern DNA, it is necessary to shear the DNA strands into small fragments to permit sequencing.

Authentication—How do you Know if it is Really Ancient DNA?

While aDNA researchers and archaeologists might do everything they can to avoid contamination, it remains a significant issue. For this reason, it is necessary to provide evidence beyond a reasonable doubt that the DNA sequences obtained from specimens are, in fact, genuine.

In analyses involving faunal remains, one can design PCR primers that do not amplify human DNA. However, possible contamination from PCR products amplified in previous studies, or the DNA of domestic animals at home, at the archaeological site, or in the general environment can still be a problem. As discussed earlier, maintenance of unidirectional workflow within the labs is designed to prevent contamination by PCR products and modern DNA. It is also worth thinking about what animals that may have shed hair and skin on-site during excavation or in the vehicle used to transport archological remains.

Ancient DNA results, like all good scientific results, must be replicable. The nature of the replication can vary depending on the novelty of the result, as well as the nature of the samples and the research question. Replication can involve two steps: internal replication and external or independent replication. All aDNA results should be internally replicable in the lab. If the specimens being studied are large enough to bc divided, at least two independent extractions should be undertaken. If the result is particularly significant or unexpected, then independent researchers in an independent laboratory should undertake the second independent extraction. For small specimens, such as rodent bones, dividing a specimen for independent extractions is unfeasible. In such a circumstance, at least two independent PCRs from an extract should be performed on different days. The products of these reactions should both be sequenced. Additionally, each strand of the DNA molecule should be sequenced, producing two DNA sequences for each PCR and a total of four sequences for each sample. Consistency between these independent tests is important for establishing the authenticity of a result. Again, if an individual result is particularly important, for example, showing that chicken bones recovered from a pre-Columbian archaeological site in South America are likely to have come from Polynesia (Storey et al. 2007), it is wise to have an external replication using different methods. As an example, one set of sequences being derived from direct sequencing and the external replication being conducted by cloning.

Ancient DNA acquires characteristic damage patterns during degradation (Binladen et al. 2006; Gilbert and Hansen et al. 2003; Gilbert and Willerslev et al. 2003; Green et al. 2006; Handt et al. 1994; Hansen et al. 2001; Hofreiter and Jaenicke et al. 2001; Höss et al. 1996; Krings et al. 1997; Stiller et al. 2006) that can be useful in distinguishing aDNA from contaminating, modern DNA. Unfortunately, damage patterns consistent with aDNA can be seen on contaminating DNA when the specimens have been stored for ten or more years after the contamination occurred (Sampietro et al. 2006). This can easily happen with archaeological remains that have been studied in museums or comparative collections. In other words, after ten years, the DNA from the excavating archaeologist or museum curator that was deposited on the archaeological bone has damage patterns indistinguishable from those of the specimen itself. Clearly, this type of contamination is particularly problematic in aDNA studies of human remains.

One of the classic methods of authentication of an aDNA sequence is whether or not, when compared with known sequences, it makes "phylogenetic sense." Put another way, does the DNA sequence that was obtained from a canid sample from an archaeological site in North America look similar to other North American canid samples? If it does not, the analyst cannot automatically reject the DNA sequence data that has been recovered. However, if the canid sequence resembles those obtained from ancient Polynesian dogs, then the onus is on the researcher to prove that there are other reasonable explanations for the finding.

What we have described is the ideal situation for extraction of aDNA. The reality is that most aDNA studies are not conducted with samples collected in this way. Studies of specimens collected in less than ideal circumstances are certainly not worthless, but the argument for the authenticity of the genetic data collected is much more difficult. Site-specific strategies to overcome the each project's particularities can be developed in collaborations between archaeologists and molecular anthropologists. The more information the molecular anthropologist has about the history and conditions of the specimens and the site from which they were recovered, the better able they will be to develop appropriate controls and strategies. Just as there is no such thing as the perfect archaeological site, there is no such thing as the perfect aDNA sample.

Ancient DNA Analysis: Applications

Before selecting bones for aDNA analysis, one must consider the question being asked and, in particular, whether DNA is the only way to obtain the information necessary to answer that question (see Box 3.1 for other considerations). Keeping in mind that aDNA analysis is destructive, expensive, and time consuming, one must assess the appropriateness of undertaking a DNA study to answer the question. Costs for DNA analysis vary dramatically and will depend on DNA preservation, the data needed, and the facilities available. Depending on the number of samples necessary to provide a clear answer, these costs tend to run in the range of thousands, if not tens of thousands, of dollars. Commercial aDNA labs are available, but these are very expensive and may not provide the necessary analytical support needed to interpret the result. So, unless the project provides the funding and direction for aDNA analyses, as a general rule, it is prudent to think about choosing specimens for DNA analysis in the same careful and limited way that specimens for radiocarbon dating

Box 3.1: The Archaeologist's Ancient DNA Field Collection Kit

Items to take into the field

1. Gloves: Take several pairs of disposable latex or nitrile gloves. Change them each time you touch a new sample (or scratch your nose). It should be noted though that some people have an allergic reaction to latex or to the talc used to powder the insides of gloves. Try them out before you go into the field or take several different kinds of gloves. Generally, the allergic reaction is merely uncomfortable and not life threatening. It would be unusual for someone to have a life-threatening allergy to latex, but be unaware of these facts. Routine visits to the doctor or dentist would typically reveal such allergies.
2. Masks: Disposable facemasks help keep your saliva off of samples you may be recovering for aDNA. This is particularly important when recovering human remains.
3. Sterile tubes: These are critical for storage of small or fragile samples. Larger samples can be stored in plastic bags or aluminum foil, but all care should be taken to make sure that these are not contaminated prior to use.
4. Bleach: Regular household bleach is ideal for decontaminating tools or cleaning surfaces.
5. Disposable plastic tweezers or chopsticks: These are necessary for manipulating bones if you are considering undertaking aDNA analyses of human remains.
6. A freezer, if possible: (Ok, you may not generally transport a freezer into the field, but you may have a handy beer fridge.) Temperature fluctuations are terrible for DNA preservation, and generally once a sample is removed from the ground (where temperatures tend to be relatively stable), the rate of degradation will increase. So, storing the samples you want to use for aDNA under a cool, constant temperature is ideal. Keep in mind, however, that keeping human remains anywhere near food or food containers is unacceptable in some cultures. Therefore, consider local customs and concerns when storing and transporting human remains.

are chosen. Project specifics may, however, provide enough funding and focus on questions that require and allow for large-scale DNA analysis, and we hope that archaeologists would consider working with aDNA researchers during the planning stages of these types of projects.

Mitochondrial DNA

The majority of aDNA studies to date have focused on mtDNA because of two main factors. Discussed at length in chapter two, mtDNA is especially useful for tracking historic relationships. Yet, the primary reason is related more to a technical issue pertaining to the degraded nature of aDNA. As also discussed in chapter two, each cell contains multiple mitochondria, and each mitochondrion contains multiple copies of the mitochondrial genome. For every copy of a nuclear gene in a cell, there are hundreds or even thousands of (depending on the cell type) copies of the mitochondrial genome. This simple numerical advantage makes mtDNA much more likely than nDNA to survive and to be recovered. In addition, mtDNA has been the focus of many modern population studies, so worldwide population databases often exist that can be particularly valuable for population-level studies of origins and affinities. This is an area where aDNA analyses of human remains can provide a diachronic perspective and make a major contribution to our understanding of prehistory.

The maternal mode of inheritance of mtDNA and the high level of variation, particularly in the hypervariable region, make it valuable for assessing maternal relatedness between individuals in a site. Biological relationships across burials can be investigated, and with this information, their associations with health and social status can be interrogated. Most studies of human aDNA have focused on the hypervariable regions and population markers such as SNPs or variations in length (indels, see chapter two). There are, however, more conserved regions of the mitochondrial genome that can be particularly useful for the identification of species. Traditional faunal analyses may not allow reliable species-level identification to be made in some cases, such as when distinguishing sheep from goats (Loreille et al. 1997). Frequently analyzed genes on the mitochondrial genome include cytochrome b (often simply referred to as Cyt b) and ribosomal subunits 12 (12s) and 16 (16s). The specific region chosen may be dependent upon the availability of DNA sequences in large, international genetic databases, such as GenBank. If such comparative data do not exist, then a comparative database may have to be generated by sampling extant species from the geographical region of interest. Recently, researchers working on bird evolution have identified one gene on the mitochondrial genome, cytochrome oxidase I (COI), that shows very little within-species variation and significant levels of between-species variation (Hebert et al. 2004). As a result, the morphological and COI-based phylogenies for the bird species show remarkable similarity. Research is ongoing to determine the usefulness of COI, and preliminary results suggest that it is useful in mammals (Francis et al. 2010; Robins et al. 2008), insects (Jung et al. 2011), fish (Kochzius et al. 2010), and a range of other animals. The analysis of variation in the mitochondrial genome of animals can be used to trace human movement and trade, investigate domestication, and contribute to the reconstruction of prehistoric diet and environment.

Nuclear Markers

If DNA is particularly well-preserved in samples, then it is sometimes possible to access nuclear DNA, and this is increasingly the case with the application of NGS. This method allows the investigation of specific gene function (Coop et al. 2008; Jaenicke-Després et al. 2003; Lalueza-Fox et al. 2008; Svensson et al. 2007), and it can identify biological sex (Cappellini et al. 2004; Cunha et al. 2000; Faerman et al. 1995; Faerman et al. 1998; Stone et al. 1996). Sex determination has probably been the most common application of ancient nDNA, and given that sex is chromosomally determined, one can either target the X or the Y chromosome for this purpose.

Molecular sex identification should be undertaken only when morphological sex determinations are problematic, such as in the case of juvenile remains, when remains are particularly fragmentary, or if the identification of sex is critical for the interpretation of the archaeological site. When molecular identifications are warranted, they are performed by targeting the X and Y chromosomes. The genes on the sex chromosomes that are useful are **amelogenin**, which differs slightly between the X and Y chromosomes and the SRY. Amelogenin is a protein that is involved in the development of enamel. One particular part of the gene that encodes for amelogenin, which is found on both the X and Y chromosomes, differs in size on each of those chromosomes. So if that part of the gene is amplified using the appropriate PCR primers, PCR amplification will produce two different-sized, PCR fragments, or two bands on a gel for an individual that had both an X and a Y chromosome, that is, a male. The PCR amplification from a female, who possesses only X chromosomes, will produce only a single-sized PCR product, or only one band on a gel. The lack of variation in the amelogenin sequences, however, means that the identification of contamination is particularly difficult. Analysis of the SRY is also used for sex determination; but, the fact that a PCR product will only be produced from a male individual makes a nonresult ambiguous. It could mean the PCR did not work because the sample came from a female, because of a lack of aDNA, or because of inhibition. Further developments in pyrosequencing for aDNA may allow for the sequencing of the full range of nuclear markers, thereby expanding the pool of questions that molecular anthropologists and archaeologists can address with aDNA.

The ability to recover aDNA sequences associated with particular functional genes is opening up new doors for aDNA research and, ultimately, for understanding prehistory. By examining changes in functional genes in populations through time, we can see evolution in action. For example, by looking at the genes for lactose tolerance in aDNA from early Neolithic farmers versus earlier hunter-gatherer in central Europe, researchers were finally able to show that genes for lactose tolerance were brought into the region with the Neolithic farmers and were not present in the earlier populations (Haak et al. 2010). This approach is also being used to help understand the process of domestication in a range of animals (discussed in chapter seven).

Ancient DNA from Other Archaeological Samples

Other types of archaeological samples also lend themselves to the recovery of aDNA. For example, artifacts that are made from organic materials, such as animal skin, plants, or bones can yield aDNA. These studies will be discussed in chapter seven. With proper methodology, coprolites, soil, plants, and microbes may yield aDNA as well.

Coprolites

DNA can be obtained from coprolites, which would contain the DNA of not only the coprolite producer but of the producer's intestinal flora and food sources. Coprolites, then, should be considered as potential sources of DNA to address many questions, including the reconstruction of past diets and environments. Once again, though, the issue of contamination due to modern environmental DNA has to be considered.

Soil DNA

Analyses of aDNA extracted from permafrost and soils in cave deposits (Willerslev et al. 2003; Willerslev et al. 2007) have indicated that plant and animal DNA can be recovered from these sources, thus allowing the reconstruction of paleoenvironments. It should be noted, however, that it is not known how rapidly and how far aDNA may migrate through soils, making chronological control problematic (Pääbo et al. 2004).

Plants

Most archaeologists deal with the remains of hard tissue, such as bones and teeth. In sites with extraordinary preservation, plant remains are sometimes recovered, and these can also be a source of aDNA (Jaenicke-Després et al. 2003). Recent studies have reported methods for the extraction and amplification of DNA from ancient wood (Deguilloux et al. 2006; Tang et al. 2011) and pollen (Bennett and Parducci 2006; Parducci et al. 2005). DNA cannot, however, be obtained from charcoal as heat destroys the DNA.

Microbial DNA

Significant efforts have been directed toward recovering DNA from infectious organisms. Claims have been made for the identification of tuberculosis (Arora et al. 2009; Barnes and Thomas 2006; Baron et al. 1996; Bathurst and Barta 2004; Crubézy et al. 2006; Donoghue et al. 1998; Donoghue et al. 2004; Donoghue et al. 2005; Donoghue et al.

2009; Fletcher and Donoghue et al. 2003, 2003b; Gernaey et al. 2001; Haas, Zink, and Molñar (nacute) et al. 2000; Matheson et al. 2009; Taylor et al. 1996; Taylor et al. 2005; Wilbur et al. 2009; Zink et al. 2003), plague (Bianucci et al. 2008; Drancourt et al. 2007; Drancourt et al. 1998; Gilbert et al. 2004; Raoult et al. 2000; Theilmann and Cate 2007; Vergnaud et al. 2007; Wiechmann and Grupe 2005; Wood et al. 2003), syphilis (Bouwman and Brown 2005; von Hunnius et al. 2007), Hansen's disease (leprosy; Donoghue et al. 2005; Haas, Zink, and Palfi et al. 2000; Matheson et al. 2009; Taylor et al. 2000; Taylor et al. 2006), flu, and even the Irish potato famine agent (May and Ristaino 2004; Ristaino 2002; Ristaino et al. 2001; Willmann 2001). Analyses of DNA from ancient pathogens will be further discussed in chapter eight.

Field Protocols

While many of the protocols available to maximize the chances of success are the responsibility of the laboratory team, there are several precautions that excavators and curators can take before specimens are submitted to an aDNA lab. The precautions suggested below are meant only as guidelines, and a quick checklist for things you should consider taking into the field with you is provided in Box 3.2. They are not substitutes for early collaboration among archaeologists, physical anthropologists, and molecular anthropologists, which will permit the development of site-specific strategies. It is worth taking the appropriate sampling equipment to the field, even if researchers don't initially intend to take samples for aDNA analysis. The process of excavation may reveal opportunities to profitably apply aDNA techniques that were unanticipated before field work began.

If at all possible, remove samples for aDNA analysis, particularly any human remains, as soon as possible after exposure. Do not wait until an entire burial has been exposed before removing samples. We recognize that this is not commonly accepted excavation practice, so the value of the potential aDNA analysis has to be weighed against the possible effect on other results. Additional aspects to consider if aDNA samples may be taken while on the site are described below, and a discussion of the ethical considerations associated with aDNA research on human remains can be found in chapter four.

Sample Collection

Whenever possible, take two bones/teeth per individual. The DNA extraction process can be undertaken from a specimen as small as a single rat femur (0.1g) when preservation is good, but teeth, phalanges,

Box 3.2: Eight Simple Steps to Doing Ancient DNA Research

1. Funding: Ancient DNA research is expensive. Consider the possibility of an aDNA component to your research before applying for funding. Establishing a collaboration with a molecular anthropologist at this stage will allow the incorporation of aDNA as part of a coherent and integrated research agenda, it will permit joint grant applications, and will prevent scrambling for money for laboratory costs after the excavations have been completed.
2. Ethics and other approvals: In addition to the approvals all excavations require, aDNA research typically necessitates bureaucratic permission to export materials out of foreign countries (if obtained overseas) and to conduct destructive analyses. In many instances, permission from local descendant communities (see discussion in chapter four) may be needed. Like permissions to excavate, granting of these approvals can take significant time, so early application is prudent.
3. Assemble a field kit for sample collection (see Box 3.1).
4. Use gloves to collect a subsample as soon as specimens are exposed, and if possible, avoid washing or handling any aDNA specimens.
5. Collect enough material for replication and the direct dating of the sample, which may be necessary depending on the genetic results obtained.
6. Keep samples in an appropriate environment. Particularly, one that is cool, climatically stable, and away from any biology laboratories.
7. If the remains for aDNA analysis are human, collect DNA samples from the field crew, remembering the need to acquire ethics approval before doing so.
8. Talk to your collaborating molecular anthropologists as early as possible.

metacarpals, or metatarsals are ideal. Unused bone can be returned at the end of the study, so it is preferable to err on the side of providing too much material. Avoid cutting or breaking specimens in the field, as this is likely to increase the chance of contamination. When the aDNA research is to focus on human remains, archaeologists should wear latex or nitrile gloves while collecting the samples. Even better when circumstances allow is for the person collecting the samples to wear the standard equipment used in the aDNA lab itself—a facemask, a hairnet, shoe covers, and a coverall in addition to the gloves. Any aDNA researcher can provide a collaborating archaeologist with these items. Subsampling of bones and other aDNA samples should be done in the aDNA lab, thus allowing the return of the unused portion of the sample to the archaeologist.

Storage Containers

Small and fragile samples, particularly of human bone or teeth, should be stored in sterile containers. DO NOT ask a biologist for sterile containers.

They will be sterile in the technical sense that they will not carry microbial life; but as with anything coming out of almost any biology laboratory, they will be covered in high-quality DNA that is perfect for contaminating ancient specimens. If such containers have not been procured before going into the field, the samples can be placed in standard archaeological ziplock bags and/or wrap them in foil. The materials should not have previously been used. Again, samples should be handled only while wearing gloves. Standard archaeological procedures for storing flora and fauna are appropriate. If samples are dry, they can be stored in sealed plastic bags, while damp samples are better stored in paper bags to prevent fungal and bacterial growth. The use of aluminum foil or aluminum containers is also acceptable but should be avoided if samples are being transported by air and likely to be subject to inspection at the airport by officials who will want to open opaque containers to see the samples.

Treatment of Bones Prior to Ancient DNA Studies

The use of x-rays and other radiographic techniques are sometimes described as nondestructive or noninvasive because they do not alter the gross morphology of archaeological remains. However, there is some evidence to suggest that x-ray procedures degrade DNA, thereby, further limiting their availability to aDNA analysis (Götherström et al. 1995; Grieshaber et al. 2008). Until more testing confirms or disproves this association, it is sensible to refrain from applying radiographic analysis to samples intended for aDNA studies. If practical, x-ray machines at the airport should be avoided. Our experience is that airport security officials can be persuaded not to x-ray samples when researchers explain that they are going to be used for aDNA studies. They typically, however, want to visually inspect those samples, so transparent storage containers are crucial. If possible, make sure that they do not open any of the bags or tubes that the samples are stored in.

Collect DNA Samples from the Field and Laboratory Crew

Even when all possible precautions are taken, some degree of contamination by modern DNA is inevitable. It is important for molecular anthropologists to be able to detect such contamination when it occurs, so they may devise a strategy to combat the problem. The approaches to overcoming the presence of contaminating DNA dramatically differ depending on whether the exogenous DNA was introduced early in the process and is present on the sample itself, or whether it was introduced during the DNA extraction process. Being able to identify the DNA sequences of the archaeologists working at a site and in the archaeological lab simplifies the problem. All that is required is a small hair sample (a few strands plucked, not cut) placed in a sealed envelope or sealed plastic bag or a buccal (cheek swab) sample. The section of the genome under investigation in the study can then be sequenced for each of the people who handled the remains and kept on file for later comparison. Ancient DNA laboratories regularly keep such comparative samples for those using the facilities.

The collection of DNA samples, even when only for comparative purposes, requires the informed consent of each individual. The management of consent procedures falls within the purview of Institutional Human Subjects Committees. It is important that institutional approval is obtained before sample collection begins, and informed consent is obtained at the time of collection. Excavating barefoot should be discouraged both because such behavior is likely to be in violation of health and safety regulations and because it poses a significant contamination risk if aDNA studies are a possibility.

Common Field Practices to Avoid

Washing the Bones

Washing bones, particularly with bare hands, facilitates the penetration of modern DNA deep into the pores of the bones. Furthermore, washing bones often accelerates DNA degradation (Gilbert et al. 2006). The most effective strategy to combat the contamination problem is to minimize the degree to which samples are exposed to modern DNA. Reducing this exposure can be achieved in the field by leaving sediment attached to the sample when it is collected. This can be removed later under controlled conditions.

Applying Glues to Stabilize the Bones

Gelatin-based glues have a long history as stabilizers for excavated hard tissues. It has been shown, however, that such glues present a significant source of contaminating DNA in ancient samples (Nicholson et al. 2002). Curated collections that have been preserved with glue should not automatically be eliminated as potential samples for study, even if it is unclear what type of glue was used. It is sometimes possible to identify the species from which the glue was made by sequencing the DNA recovered following amplification with general mammalian PCR primers. Then, if there is sufficient phylogenetic distance between the glue species and the target species, highly specific PCR primers can be designed that preferentially bind to the DNA of the target species.

Spitting on or Licking the Bones

By now, this should be obvious; DON'T! We will not mention this again.

Chapter Four

Ethics of Molecular Anthropological Research

Archaeological and genetic studies of human prehistory confront ethical issues with respect to the construction of historically meaningful interpretations and the dissemination of results to the communities who may be affected by them.

Here, we present a picture of the development of the Human Genome Project (HGP) and the Human Genome Diversity Project (HGDP) because their histories are illustrative of the way that ethical issues have been confronted as they have arisen in studies of modern DNA samples. This review also illustrates why many indigenous communities are suspicious of genetic studies, and it highlights how some of the ethical issues raised around the topic of documenting genetic variation in human populations are seen in very different ways by some members of the scientific and indigenous communities. We follow this review with a discussion of the additional issues associated with analyzing DNA preserved in archaeological remains.

The Human Genome Project

In October, 1990, the HGP was officially launched. Initially planned as a fifteen-year project involving researchers from eighteen countries, technological advances resulted in an early completion of the project, with the first draft of the human genome sequenced in 2001. The Human Genome Organization (HUGO) coordinated the international effort and continues to drive research today. The project goals were to construct physical maps of gene locations, sequence the three billion bases of human DNA, and identify all human genes. The goal then was to make

---

DNA for Archaeologists by Elizabeth Matisoo-Smith and K. Ann Horsburgh, 81–95.

that sequence information available and accessible for further study. A final goal was to further development of DNA sequencing technologies. The project was initially funded by the United States Department of Energy and the NIH with a total budget of US$437 million.

The HGP is generally heralded as a major scientific success story—it was completed under budget and ahead of schedule. Part of this achievement, no doubt, was the result of the development of a direct challenge to the HGP when Craig Venter established, Celera Genomics and stated that he would be the one to sequence the complete human genome. In February, 2001, the two teams published draft sequences; the HGP and collaborators published in *Nature*, and Venter and his colleagues in *Science* (International Human Genome Mapping Consortium 2001; Venter et al. 2001). The HGP was declared finished in 2003 with their publications of the full sequence.

There is still much that we do not know about the human genome, and research to provide these insights is ongoing. According to the Oak Ridge National Laboratory's HGP Information site (http://www.ornl.gov/sci/techresources/Human_Genome/project/journals/insights.shtml), sequencing the entire human genome revealed that it is 3.2 billion nucleotides in length. The average gene is about 3 kilobases long, but the largest is 2.4 million bp long. Most of the genome (over 50%) is made up of repeat sequences, which do not code for any proteins. It has been estimated that there are about 25,000 genes, or those sections that code for proteins, in the total genome. This is far fewer than most of the researchers expected prior to undertaking the project. We understand the function of fewer than 50% of those genes. Even today, over twenty years after the launch of the HGP, we still have a long way to go before we fully understand the full structure and function of the human genome.

While formal international collaborations inevitably result in some tension between governmental and scientific goals, when private funding gets added to the picture, competing interests become even more complicated. This is indeed what happened with the HGP. The whole concept of international collaboration and information sharing was doomed when patents and the protection of trade secrets became an issue. Big business sees opportunities for investment, targeted drug development, and other therapies with potential pharmaceutical spin-offs. Governments are particularly interested in technology and job creation. In large science projects like this, one might wonder how much thought is given to the individuals whose genes are being mapped and whose DNA is being sequenced. For researchers focused on the value of the knowledge obtained and the potential for future treatment of diseases, the concerns of and for the individual sometimes seem secondary. The individuals providing the samples, however, may calculate these priorities very differently.

Interestingly, from its conception, the HGP had a major focus on the ethical, legal, and social implications (ELSI) of the research. A portion of the annual budget was made available to fund discussion and research on ELSI issues, and an ELSI working group was established. Yet, by 1992, a report written by the House of Representatives pointed out that "there is no existing policy process that will use the results of the ELSI research to make recommendations" (McCain 2002:117). Basically the ELSI programs had no teeth. One of the biggest concerns of many people when it came to the HGP was the commercialization of genetic information and technologies, and the ELSI working group did little to address this issue. This is a valid complaint, as the United States Patent and Trademark Office has granted over a thousand patents on human DNA. Patent protection of genes or gene sequences are said to restrict consumer access to genetic technologies and prevent commercial labs from offering genetic tests to patients. In an analysis of the effectiveness of what was supposed to be the HGP's conscience, the ability of public science programs to monitor the ethical and social impacts of their own work was questioned, with reviewers arguing that the ELSI programs had been "falling short of their commitment to act in the public interest" (McCain 2002:131–132).

The genome sequence that the HGP constructed does not represent the DNA of a single individual (though the sequence created by Venter's group is thought to be his genome). It is instead an amalgam, or reference genome of many individuals, both male and female. One of the major debates that arose early in the HGP pertained to how representative of the human species a single genome might really be. While it has been claimed that all human DNA is 99.9% similar, we also know that humans and chimpanzees share a 98.4% similarity in their DNA (Mikkelsen et al. 2005). Rarely do we have to argue that the phenotypic, or physical, differences between humans and chimpanzees are significant. Therefore, a group of population geneticists and evolutionary biologists led by Luca Cavalli-Sforza, from Stanford University, contended that the HGP should at least consider human variation (Cavalli-Sforza et al. 1991).

In a paper presented in the journal *Genomics*, Cavalli-Sforza et al. (1991:490) pleaded with geneticists and public and private agencies to "grasp a vanishing opportunity to preserve the record of our genetic heritage" before it was "irretrievably lost." They went on to suggest that "the populations that can tell us most about our evolutionary past are those that have been isolated for some time, are likely to be culturally and linguistically distinct, and are often surrounded by geographic barriers. . . . [because] [i]solated human populations contain much more informative genetic records than more recent, urban ones." (1991:490).

They argued that the collection of these samples must "entail a systematic, international effort to select populations of special interest throughout the world, to obtain samples, to analyze DNA with current technologies, and to preserve samples for analysis in the future" (1991:490). The authors recognized the ethical issues that would be raised in the study.

> Among these very informative groups have been many peoples historically vulnerable to exploitation by outsiders. Hence, asking for samples alone, without consideration of a population's needs for medical treatment and other benefits, will inevitably lead to the same sense of exploitation and abandonment experienced by the survivors of Hiroshima and Nagasaki. It will be essential to integrate the study of peoples with response to their related needs. (Cavalli-Sforza et al. 1991:490)

The Human Genome Diversity Project

Following the plea of Cavalli-Sforza and colleagues, HUGO formed a committee to investigate the possibility of a major study of genetic variation. After much discussion, the HGDP, a subproject of the HGP, was formed in 1992. Funding for the planning process was provided by the NSF, the National Human Genome Research Center, the National Institute of General Medical Sciences, and the United States Department of Energy. The organizations planned three meetings to start discussions and planning. The first meeting was to address the issue of sampling strategies. Later meetings were to focus on the selection criteria for which populations should be targeted. A final meeting was to deal with ethical and technological issues associated with such a project.

The original organizers of the HGDP—Cavalli-Sforza, Allan Wilson, from the University of California, and others—were not only interested in the structure and function of the genome, they had historical interests in the evolutionary history of the human species and the development of the diversity seen in modern humans. While the HGP was focused on medical questions, the HGDP was to focus on the intellectual benefits of understanding human genetic diversity and its origins. However, even before the first meeting was held, one of the major stumbling blocks for the project became obvious. Cavalli-Sforza and Wilson had taken very different approaches in their previous studies of human diversity. Cavalli-Sforza focused on population variation, and Wilson focused his research at the individual level. These differences influenced their opinions about how the HGDP should sample human variation, which became a major debate in the project as a whole (Reardon 2001).

Whether diversity is organized at the individual or group level is a classic debate in genetics. One of the complex issues about assessment of group diversity is how one defines the group—by race or by population

(Dobzhansky 1937). This biological question about group diversity thus feeds into the social debates about the concepts of race, with all of the associated historical baggage. After World War II, biologists and biological anthropologists were adamant that there is no biological validity to racial classification (Biondi and Rickards 2007; Brace 2005; Marks 1995). Even the use of the term race is frowned upon because of the historic associations and social implications. We now find that the words "population" or "ethnicity" tend to be used in place of race, and genetic studies have shown that there is as much variation within a population as between populations (AAPA 1996). However, with increasing genetic data on population variation and, in particular, the growth in the field of medical genetics, debate regarding the validity of the term is returning (Gravlee 2009; Hunley et al. 2009). But how does one define human populations? Are human populations definable biologically, or are they social constructs?

At the first meeting of the HGDP, held at Stanford University in July, 1992, forty researchers, including physical and cultural anthropologists and population geneticists, attempted to design a sampling strategy. Unfortunately, Allan Wilson was not among those at the meeting, as he had passed away on the 21st of July, 1991, but, his former students, Mark Stoneking, Mary-Claire King, and other close colleagues were there to pursue his vision. Several issues were raised and debated during the three-day meeting. The first was the basic sampling procedure.

The population approaches supported by Cavalli-Sforza proceeded to identify appropriate populations from which to sample genetic variation. These groups were identified based on a number of anthropological criteria and other factors, such as geographic isolation. Language was identified as a key indicator of group identity. The proposal was to collect blood from fifty individuals each from two hundred aboriginal tribes defined according to language. Aboriginal tribes were identified because they would be representative of biological variation prior to the expansion of present-day, dominant groups. A date of 1492 was chosen to define the point at which this displacement of aboriginal groups was to have begun.

Mark Stoneking raised his concerns regarding the criteria for defining the aboriginal groups to be sampled. He argued that the population approach was based on numerous assumptions because "it just focused on well-defined ethnic and linguistic groups. And when you're done with your survey you will find that the human species is made up of well-defined ethnic and linguistic groups. By sampling that way you bias the results" (Roberts 1992a:1205). Similar concerns were voiced about the lack of recognition of the significant expansions of populations over the thousands of years prior to the arrival of Columbus in the New World.

Wilson's individual approach recommended a geographic grid strategy in which fifty individuals would be sampled along evenly spaced locations around the world. Using this approach, it was argued, would allow scientists to test whether or not genetic variation was indeed correlated with linguistic or cultural variation. Cavalli-Sforza was critical of this approach, claiming that it was impractical and unlikely to sample the necessary aboriginal communities.

In addition to debate over the sampling strategy, other issues were also raised. The proposed numbers of samples needed for each population and the type of samples collected were also discussed. Cavalli-Sforza's goal of a minimum of fifty samples per population was questioned, to which he was said to reply, "One person can bleed fifty people and get to an airport in one day" (Roberts 1992a:1205). So, it seemed that logistical rather than scientific issues were driving the proposed sampling strategy. The biostatisticians present at the meeting pointed out that smaller sample sizes would be adequate if the number of genetic markers being studied was increased. They suggested that for most questions the group was interested in, sample sizes as small as twenty-five or even ten individuals would suffice (Roberts 1992a).

The proposal to establish permanent cell lines from samples collected from aboriginal communities was another major contentious issue for the HGDP. If DNA is extracted from blood or bone, there is a finite amount, allowing for only a restricted number of experiments. If cells are cultured, or grown, in the laboratory, one can take advantage of the replication activity inherent in living cells. However, once removed from the body, cells will only replicate a limited number of times before cell death occurs. White blood cells, if transformed by infection with a particular virus (the Epstein-Barr virus), become "immortal." That is, they are no longer subject to the normally programmed cell death and can be grown and stored indefinitely.

One of the goals of both the HGP and HGDP was the production of immortal cell lines to provide an endless resource for scientists. Logistically, the production of cell lines is difficult because blood samples generally need to reach a lab within forty-eight hours of collection to ensure viable white blood cells for transformation. Given the practical and financial constraints associated with cell line production (it was estimated that each sample might cost up to US$500 to produce), it was suggested that "cells from only the exceptionally interesting and unique populations need to be immortalized; for other populations, only blood samples need to be collected, and the DNA extracted from these" (Bowcock and Cavalli-Sforza 1991:496). Once again, the Wilson group suggested collecting hair or cheek cell samples, which would be cheaper, less invasive, and would provide enough DNA for up to 1,000 PCR experiments. This amount was considered to be enough to answer the key questions that were being addressed by the project. In addition, this approach would cost between US$5 and $10 per sample to collect and process. The desire for the production of cell lines, however, would turn out to be much more than just a logistical and financial debate. It was soon to be the topic of major discussions relating to ethical issues associated with informed consent and intellectual property.

The next HGDP meeting, held at Penn State in October, 1992, brought together fifty archaeologists, linguists, and other anthropologists to decide which populations would be included in the study. Descriptions of the meeting suggest that there were two very distinct camps—those familiar with biology and those familiar with culture. There was very little overlap. The meeting was broken up into regional committees who all had different approaches and different criteria for defining their populations of interest. Some groups focused on linguistic diversity, others on phenotypic diversity, and still others on isolation and representativeness. At the end of the meeting, lists identifying populations of interest, anthropologists to contact, and blood samples already held in freezers around the world were collated. There was, however, no ultimate consensus about the sampling strategies, or central purpose of the HGDP itself (Roberts 1992b). It was apparent at the end of the Penn State meeting that the different regional criteria for group definition were in conflict. The list of defined populations was discarded, and in Sardinia in September, 1993, a HUGO workshop was held to further discuss the categories of populations.

It was not until a third meeting was held in Bethesda, Maryland, in February of 1993 that the ethical, social, and legal implications of the HGDP were to be addressed, which was a move prompted by concern voiced by the National Institutes of Health. It is unfortunate that not only were representatives of indigenous communities not invited to the Bethesda meeting, but that indigenous leaders perceived they had been explicitly excluded from discussions because the *scientific* nature of the project required input only from *scientists*. Leaders of HGDP argued that indigenous groups misunderstood and were misinformed particularly in regards to the commercialization of samples and the claims that sampling had already begun (Cavalli-Sforza 2005; Reardon 2001).

In the spring and summer of 1993, the HGDP began to implode. After the Bethesda meeting project, leaders were contacted by the Rural Advancement Foundation International (RAFI), a Canadian-based organization dealing with biodiversity and intellectual property rights. They were concerned about the possible exploitation of marginalized communities for financial gain, the patenting of genetic diversity by industrialized countries, and the possible diversion of funds into HGDP coffers for development. Additionally, the warehousing of genetic samples, such as transformed cell lines, prompted concerns that this might allow pharmaceutical companies to develop group-specific bio-weaponry (Kahn 1994). Soon after, several indigenous rights groups started discussions about how the HGDP might impact the construction of indigenous identity. Indigenous community networks began calling for organized protests in response to the HGP and HGDP.

> What appears to be an innocuous proposal to further science is in reality a proposal which can set a precedent to undermine Native American policies that protect tribal sovereignty and self-determination. Research predicated on the assumption that *they* and not we know what is best for our communities as well as a reluctance to be forthright reinforces the continuation of paternalistic and colonizing attitudes toward indigenous communities. (Lone Dog 1999:63)

Legitimate concerns were raised by indigenous rights groups because of well-known historical abuses of biological data, but more recent unethical behavior by anthropologists indicates that those lessons had not been learned (Wiwchar 2004). For example, human geneticist Ryk Ward and colleagues collected 883 samples from Nuu-chah-nulth people of the Canadian Pacific Northwest Coast between 1983 and 1985 for a study of rheumatic disease (Atkins et al. 1988). Informed consent was granted specifically for the rheumatic disease research. It was the largest ever study of genetic markers among Canadian First Nations peoples. The original researchers, as well as others to whom they were inappropriately loaned, used the DNA samples in studies addressing a range of issues, including the origins of the tribe (Ward et al. 1991; Ward et al. 1993), without anyone having obtained further informed consent for other studies. The news of this misuse of samples became public knowledge through the Canadian press in 2000. In 2004, the DNA samples were returned to the University of British Columbia but not before the reputations of molecular anthropologists were substantially tarnished. Native advocate D. Wiwchar of the Nuu-chah-nulth stated that "unfortunately, Ward taught us not to trust researchers just like the Residential School taught us not to trust the Church" (Wiwchar 2004:4).

Although, federal funding had been provided for the planning stages of the HGDP, it had not yet been secured for the sampling and analysis. One of the final decisions made at the Bethesda meeting was that yet another meeting should be held, this time inviting peoples from developing countries around the world to once and for all discuss the politics of their participation in the HGDP and how the project should be run. That meeting was held in Sardinia, Italy, in September, 1993. The organization compiled a summary document that provided an overview of all of the previous meetings. Members of the group then drafted a formal proposal, which they presented to the Council of HUGO in 1994. The council approved the proposal that same year. Four main points relating to the value of the proposed project were presented: 1) the project had "enormous potential for illuminating our understanding of human history and identity;" 2) it would "provide valuable information on the role played by genetic factors in the predisposition or resistance to disease;" 3) it would link biologists with researchers in the social sciences; 4) it would thus create a "unique bridge between science and the humanities;" and 5) the project would finally "help to combat the widespread popular fear and ignorance of human genetics and [would] make a significant contribution to the elimination of racism" (Human Genome Diversity Project 1994:5).

As a result of the Sardinia meeting, the HGDP recognized the need for a focus on the ethics and politics of the project and created a North American regional committee (NAmC). The NAmC applied for funds from the MacArthur Foundation to write a model ethical protocol. The MacArthur Foundation provided the funds, but they required the NAmC to consult with Native American activists whom they would choose. That consultation took place in the spring of 1994 in San Francisco, and it did result in a model ethical protocol for collecting DNA samples. The protocol addressed several key issues including privacy and confidentiality. However, one of the major contributions it made was the suggestion that, in addition to individual consent, researchers were to seek group consent from participating communities. The process of obtaining group consent is fraught with practical and ethical considerations of its own, not least of which are determining the membership of a group or identifying who has the right to speak for a community. In addition, there is always the possibility of the rights of individuals being overridden by the rights of the identified authorities of the group. Despite these issues, the concept of obtaining group consent, the recognition that there might be different ways of working with different communities, and that the communities themselves might be the most appropriate entities to make those decisions was generally seen as a positive step.

Both the protocol and the summary document that resulted from the Sardinia workshop were presented to the NSF and the NIH who consulted widely with various public groups and experts in a range of fields. Unfortunately, the public reaction of many indigenous communities to all of the previous discussions meant that obtaining any positive feedback would be difficult. Controversy over the proposal was international. Indigenous communities felt they were being treated like lab animals, and despite the attempts to be seen otherwise, the HGDP's behavior was seen as a demonstration of "genetic colonialism" (Kahn 1994:720). In reaction to the large amount of debate over the proposed project, the NSF and NIH asked the National Research Council (NRC) to evaluate the proposal and help them make a decision regarding the future of the HGDP.

In 1997, the NRC released its report "Evaluating Human Genetic Diversity" (NRC 1997). The report claimed the "precise nature of the proposed HGDP was elusive" in intent and structure (NRC 1997:1). Because there was no clear proposal to really assess, the review committee decided that they would only consider the organizational, policy, and ethical issues that any proposed genetic diversity study would face. The report presented a number of different strategies for moving ahead with a diversity project, and one of those strategies was to abandon the biomedical focus. A project focused solely on studying diversity to better understand human history was seen as having far fewer ethical and legal complications. Much discussion was given to sampling protocols highlighting the concern for individual and group privacy and the need for consent both at the individual and group level. The committee suggested that blood samples collected should primarily be turned into purified DNA as opposed to immortalized cell lines. It is unclear whether this decision was due to the logistical problems and costs of cell line generation or to the concerns of indigenous communities over the ethics of producing unlimited amounts of their DNA, which could be used in future studies that may not have been discussed or even possible at the time of collection. It was also pointed out that "[i]t is not ethically or legally acceptable to ask research participants to 'consent' to future but yet-unknown uses of their identifiable DNA samples" (NRC 1997:8). As a result, the committee recommended that samples should not be traceable back to the individual and, if possible, not back to the specific group from which they were collected. This begged the question whether it was permissible to use samples for research projects that were not consented to as long as those samples could not be tracked back to the donor. A comment from one representative of the World Council on Indigenous Peoples indicates that this logic is flawed: "The assumption that indigenous people will disappear and their cells will continue helping science for decades is very abhorrent to us" (Kahn 1994:721).

In the end, the NRC report basically suggested that a project like the HGDP would be valuable, but it would need to be an international effort, and given the likely costs, such a global effort could not be fully funded by the NSF and NIH. Instead, they recommended that funding might be made available for United States-based projects and that the project could then be expanded only after the United States-based study was well-underway. They also identified the fact that numerous DNA or blood samples may already exist in collections around the world, and that many of these could possibly be available for the HGDP study. They recommended that the NSF or the NIH should try to find out if this were indeed the case.

Ultimately, however, little to no funding has been made available for new sample collection. Researchers involved in the HGDP and others that they have contacted have pooled their previously collected samples and cell lines, and just over 1,000 of these are now held in a repository at the Centre for the Study of Human Polymorphisms (CEPH) in Paris. Aliquots of the DNA obtained from these samples, representing at least fifty-two populations from around the world, are made available, free of charge, to any researcher conducting nonprofit research (Cavalli-Sforza 2005).

The Genographic Project

It was suggested that it could take up to ten years for the furor over the proposed HGDP to die down within indigenous communities (Kahn 1994). Given this prediction, it is somewhat ironic that just over ten years from the date of the released consensus report on the HGDP, a new genetic diversity project was conceived. In 2005, the Genographic Project was launched. The project is a collaboration between National Geographic and IBM, with funding for the fieldwork provided by the Waitt Family Foundation. The project, which was the brainchild of National Geographic Explorer in Residence, Spencer Wells, was initially developed to be an international, five-year project, combining the use of cutting-edge genetic and computational technologies in what would be "a landmark study of the human journey." The Genographic Project's visionary goals are outlined below.

> Three main components of the project are: to gather field research data in collaboration with indigenous and traditional peoples around the world; to invite the general public to join the project by purchasing the Genographic Project Public Participation Kit, and to use proceeds from the Genographic Project Public Participation Kit sales to further field research and the Genographic Legacy Fund, which in turn supports indigenous conservation and revitalization projects. (https://genographic.nationalgeographic.com/genographic/index.html)

Many critics suggest that the Genographic Project, now in its seventh year, is merely the reincarnation of the HGDP. However, there are some significant differences. The Genographic Project is clearly and adamantly nonmedical, focusing solely on mitochondrial and Y chromosome data to investigate only population and human history. Further, no immortal cell lines are being created, and no patents will be sought as a result of the research. DNA samples stay in the region in which they were collected and are not being warehoused in any central location. Individual participants have the ongoing right to withdraw or modify their consent, and they have control over the eventual storage, return, or destruction of the DNA they provided. They are also able to decide if, and how, results of the research are returned to the community. Each of the twelve regional investigators—including the Pacific region, for which author Matisoo-Smith is currently the principal investigator—obtains local Institutional Review Board approval, as well as necessary national and local permissions. Communities and individuals have a broad range of options to choose from in the donation of samples, and in most cases only buccal swabs or mouth wash samples are being collected. The Genographic Project advisory board includes indigenous advocates, geneticists, cultural anthropologists, linguists, archaeologists, and the chair is a professional ethicist. Much has been learned since and perhaps as a result of the HGDP proposal and resulting discussions and debates. Improved communication and knowledge among both scientists and communities has initiated important conversations and awareness. Both sides know better what kinds of questions to ask each other, and when to ask those questions, which is preferably in the early stages of research design. Community-based ethics review committees are increasingly common, and we hope that most researchers have learned from the difficulties confronted by earlier studies.

Despite the fact that the Genographic Project seems to have overcome many of the issues that sank the HGDP, it has not been without its own problems. Some samples collected from Native American communities early in the project were returned after local review committees decided that there was not enough detail provided on the participant information sheets (Wells and Schurr 2009). However, research is now proceeding again in North America, as well as around the rest of the world. Several collaborative projects involving indigenous communities including studies of genetic variation in the Seaconke Wampanoag tribe in Massachusettes (Zhadanov et al. 2010), the Native American descent community in Bermuda (Gaieski et al. 2011), and several tribes in the Northwest Territories of Canada and southeast Alaska (Owings et al. 2011; Schurr et al. 2011) have been undertaken, and many more are currently underway. While it was initially expected that the project would run only for five years, additional funding has been secured to allow continuation through to the end of 2012, and the possibility of future funding is being investigated.

Obtaining Consent and Other Ethical Issues Associated with Ancient DNA

Clearly, the issue of informed consent is complex when dealing with genetic studies of modern communities. Concerns get even more complicated when we consider aDNA. How does one obtain consent to take DNA from an ancient or archaeological sample? This may bring about the question of ownership. Certainly, if one wants to obtain DNA from a sample held in a museum, then permission is generally first obtained from the appropriate museum personnel. This may require a full application to a museum committee or even an application to a national organization, such as a National Heritage Council. Is this enough? What about obtaining consent from the descendants of the sample in question? Cultures and people are not, of course, static entities—so how do we then define who the descendents are? This was one of the major debates in the Kennewick Man saga. There are no easy answers.

When excavating a site and working with indigenous communities or land owners, it is most important that fully informed consent for any possible DNA study is obtained during the initial stages. This should include a discussion about the methods used, the possible results, and their interpretation and dissemination. Not only should these discussions involve what can be done with the DNA data, but also what cannot be done.

There are also key issues that need to be addressed regarding the interpretation of the data, such as who will have access to the information and to any remaining DNA. How will the DNA be disposed of once the study is finished? Can it be stored for future use? If so, who should have access to that material? Another question often asked is if scientists can recreate an organism if they have the DNA sequence. One of the requirements for publication of DNA sequence data in most of the reputable scientific journals is that the DNA sequences must be deposited in an international, public database, such as GenBank. This requirement needs to be discussed with the communities before any sample is processed. In our experience, once people have come to understand what can and cannot be done with a printed version of a DNA sequence, they have agreed to this requirement. In good conscience, however, a researcher has to be willing to accept that a community may not agree to the inclusion of genetic information about their ancestors or other culturally significant information in a public database. In such cases, this may mean that the research does not go forward.

Ancient DNA Analyses are Destructive

One of the major issues of dealing with samples for aDNA is the destructive nature of the extraction methods. Certainly, as methods improve, less and less material is required, but the DNA extraction process will always have some impact on the sample. Given this feature of DNA work, the benefits of the possible data generated must, therefore, outweigh the loss of the sample or a portion of a sample. Of course, sometimes the DNA is too degraded, and it is impossible to get any result at all. Again, this needs to be made clear at the beginning of the study to all involved.

Interpretation of Ancient DNA Results—Definitions

Archaeologists and prehistorians often incorporate the information from molecular anthropology in their syntheses and explanations of particular regions or the spread of particular archaeologically identifiable populations. How genetic variation is described, which often includes definitions or descriptions of population-specific, genetic markers, inherently defines populations that may be ancestral to the populations living in the region today, but who may not necessarily be defined in the same way. For example, in the Pacific region, a particular combination of mtDNA mutations has been identified and is regularly referred to as the "Polynesian motif" (Melton et al. 1995). The spread of the Polynesian motif has been linked to the spread of a particular cultural complex, the Lapita cultural complex and has, therefore, been tied to an archaeological population. The distribution of both the so-called Polynesian motif and Lapita spans Polynesia and Melanesia. Again, population labels that may have no biological reality can carry great social significance. To inform a self-identified and proud Melanesian that they are carrying a genetic marker known as the Polynesian motif may have significant implications for that person's identity and, thus, potentially for your relationship with that individual or that community.

Genetic information may also carry with it implications of aboriginality when particular population genetic markers are linked to population expansions that are chronologically defined. What are the implications of telling a member of an aboriginal community that they are carrying markers associated with later population movements? This information, if not fully understood and explained to individuals and communities, could have an impact on land claims, for example. Does carrying a European Y chromosome, which may be a marker of an event that happened hundreds of years ago, make one less indigenous? As scientists, we can argue that it does not, but we must consider the impact that this kind of information may have on individuals and/or communities that are not familiar with inheritance patterns of genetic markers. In this case, it is very important to make sure that when working with communities and incorporating genetic information in narratives of history, that the time and effort is made to explain what the DNA evidence means and what it does not mean. It is also important to realize that these issues are not exclusive to the study of human genetic variants, because in some cases, animals or plants may also carry significant cultural value. In many cases, these issues are as real and important to local communities as in studies of human genetic variation.

Many researchers see the ethics requirements of community consultation as a box to tick, or merely another hurdle to overcome before research can commence. In our experience, community involvement in the genetic study of modern and ancient samples has been a positive experience and has contributed in numerous ways from the very conception of the research questions that we ask to the interpretation, presentation, and interest in final results. In some cases, it has led to significant future projects.

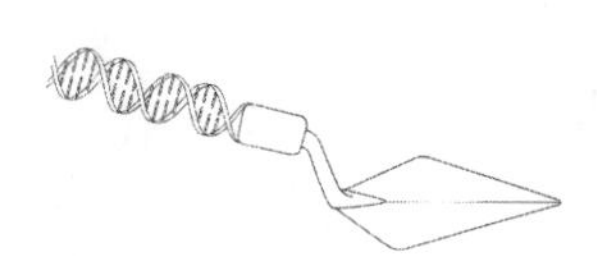

Chapter Five

Hominin Origins and Relationships

Vincent Sarich and Allan Wilson first addressed molecular evolution at the protein level in the hominin (then hominid) line in their analyses of serum albumin variation in humans and a range of other primates (Sarich and Wilson 1967b). As technologies developed, Sarich, Wilson, and other researchers studying the similarities and differences at the DNA level found that the difference between humans and chimpanzees was very small indeed—and only in the range of a few percent (Hoyer et al. 1972; King and Wilson 1975; Sarich and Cronin 1976; Sibley and Ahlquist 1984). Since those early days of molecular anthropology, the fossil record and further molecular research have not materially altered this view. It has also been shown that humans and chimpanzees are more closely related to each other than either are to the other great apes (Ruvolo et al. 1991). The date of divergence, initially estimated by Sarich and Wilson as 5 MYA (Sarich and Wilson 1967a), is still within the range of 5–7 MYA as suggested by both the fossil and the molecular evidence (Bradley 2008 and references within).

Comparison—The Human and Chimpanzee Genomes

In 2001, the first complete human genome was published (Venter et al. 2001), and in 2005, the Chimpanzee Sequencing and Analysis Consortium published a draft sequence of the common chimpanzee (*Pan troglotytes*; Mikkelsen et al. 2005). The study of the chimpanzee genome revealed that the two genomes differ by only 1.2%. Of particular interest is their identification of the gene regions where the human and chimpanzee genomes differ, as these regions might provide information associated with significant events in human evolution (Mikkelsen et al. 2005).

DNA for Archaeologists by Elizabeth Matisoo-Smith and K. Ann Horsburgh, 97–107.

Human-specific changes have occurred in parts of the genome associated with cell death in the brain, which may provide insight for the unique human condition of Alzheimer's disease. In addition, human-specific changes have been identified in gene regions associated with inflammatory response and parasite resistance. Interestingly, significant changes in the chimpanzee genome are found in areas associated with several human disease variants, including coronary artery disease and diabetes mellitus, which indicate that the human variant is very possibly the ancestral state. Further, several regions of the human genome show an unusual lack of diversity compared with the chimpanzee genome, suggesting that humans have probably experienced selective sweeps within the last 250,000 years (Mikkelsen et al. 2005). In other words, something in these regions is likely to have provided a significant advantage to the carriers so that the genotype seen today is unusually homogenous.

Most of these nonvariant parts of the genome have not yet been associated with any known functional elements, but they will no doubt be candidates for further research. One of these regions, however, is located on chromosome 7 and has been linked with two genes that we know something about—FOXP2 and CFTR. The FOXP2 gene is a regulatory gene that is somehow involved in modulating the plasticity of neural circuits in the brain. It is sometimes referred to as the "language gene" because mutations in this gene cause severe speech and language disorders in humans (Enard et al. 2002). It was initially thought that the changes seen in humans might be related to the attainment of fully grammatical language and the fine motor control necessary for the physical production of speech in modern humans. Yet, it has since been shown that the acquisition of language and speech is clearly much more complex (Vargha-Khadem et al. 2005). The gene, FOXP2, is also implicated in lung development. Located near the FOXP2 gene on chromosome 7 is the CFTR gene. Mutations in this gene lead to the development of cystic fibrosis. The comparisons between human and chimpanzee sequences indicate that the CFTR gene was a target of selection in European populations. It has been suggested that individuals who are heterozygous for the CFTR mutations (meaning that they only have one functioning copy of the gene) are less severely impacted by typhoid, cholera, tuberculosis, and diarrhea (Gabriel et al. 1994; Modiano et al. 2007; Pier et al. 1998; Williams 2006).

Although one of the earliest applications of molecular anthropology, which addressed the evolutionary relationships between humans and other primates, was responsible for the removal of *Ramapithecus* from human ancestry, it was not until very recently that improved aDNA methods allowed us to obtain DNA data that directly relates to hominin remains. These new results are providing important insights to hominin evolution. Ancient DNA researchers are often asked if one can get DNA out of fossil remains. Of course, technically, once a bone is fossilized, it no longer contains organic material and, therefore, it will not contain DNA. DNA can, however, be obtained from ancient remains, including those of extinct species.

Ancient DNA—How Ancient is Ancient?

So what is the oldest DNA yet recovered? While claims have been made for the recovery of DNA from dinosaur remains (Woodward et al. 1994) and from insects preserved in ancient amber (a la Jurassic Park) (Cano and Borucki 1995; Cano et al. 1993; DeSalle et al. 1992), these results have either never been replicated or have been proven to be false (Pääbo and Wilson 1991; Pääbo et al. 2004; Zischler et al. 1995). In most cases, the sequences obtained have been shown to be the result of contamination. Ancient plant and insect DNA has been obtained from ice-cores in Greenland, which date to as old as 800,000 BP (Pääbo et al. 2004; Willerslev et al. 2007), but such preservation conditions are extremely rare. The oldest skeletal remains from which aDNA has been recovered are dated to the tens of thousands of years before present, and these are also found in either permafrost conditions or in relatively cool and dry cave deposits (Barnett et al. 2009; Hofreiter et al. 2002; Hofreiter et al. 2007; Knapp et al. 2009; Reed et al. 2003; Smith et al. 2003). However, with constantly improving techniques and NGS, researchers are able to get DNA out of older and older samples, allowing for direct comparison between modern humans, ancient humans, and extinct hominins as long as organic material is preserved.

Neanderthal DNA

In 1997, Krings and colleagues published the first DNA sequences obtained from a Neanderthal. Their paper describes how the team was able to sequence a small portion of the hypervariable control region of mtDNA from the Feldhofer Cave specimen. This was the original type specimen discovered in the Neander valley of Germany in 1856 and dated to between 30 and 100 KYA. The analysis of approximately 360 bp of mtDNA identified an average of 27.2 SNPs (with a range of 22–36 SNPs) between the Neanderthal sample and 2,051 modern human sequences. The modern human samples differed from each other on average by 8 bp, and by a maximum of 24 bp. They differed by an average of 55 bp from homologous chimpanzee mtDNA sequences. The Neanderthal sequence was thus conspicuously closer to the sequences of modern humans than it was to those of chimpanzees, but it was outside the range of variation seen within modern humans. In addition, when compared with a worldwide sample of modern human mtDNA, the Neanderthal sequence was no closer to European populations than it was to any other group. This was interpreted to strongly suggest that Neanderthals were a separate species of hominin, rather than a subspecies of *Homo sapiens* (Krings et al. 1997).

Several questions, however, were raised about this interpretation. Given that the Neanderthal sample was at least 30,000 years old, it might be expected that it would not fit within the variation seen in living humans even if it were a member of our species. In addition, it was only one sample. To address this, several other partial mtDNA sequences of Neanderthal remains were obtained over the next few years (Caramelli et al. 2006; Gutiérrez et al. 2002; Krings et al. 1999; Ovchinnikov et al. 2000). All of the mtDNA sequences obtained from Neanderthals resulted in the same pattern in that they, as a group, fell outside the range of variation seen in living humans. In 2008, a complete mitochondrial genome dating to 38 KYA was recovered from a Neanderthal excavated at Vindija Cave, Croatia (Green et al. 2008). This was followed by four more complete Neanderthal mitochondrial genomes dating to between 38 and 70 KYA. Two of the individuals were from Feldhofer Cave, Germany, one of which was the Neanderthal type specimen (~40 KYA). One was from El Sidron Cave, Spain (~39 KYA), and one was from Mezmaiskaya Cave, Russia (~60–70 KYA) (Briggs et al. 2009). The results of these analyses were consistent with those previously found in which Neanderthals form a monophyletic clade distinct from that of modern humans. The sequence obtained from the oldest and most easterly Neanderthal from Mezmaiskaya Cave was also the most divergent from the rest of the sample, suggesting that Neanderthals exhibited some limited population substructure across time or space. Although, other indications are that mtDNA variability was not substantial. Two individuals sharing identical mitochondrial genomes were found in Croatia (Vindija Cave) and Germany (Feldhofer), as well as in two individuals both from Vindija Cave but separated by some 6,000 years (Briggs et al. 2009). To really address the issue of how Neanderthals were related to anatomically modern humans, however, it was necessary to compare the mtDNA of the Neanderthals with a sequence obtained from similarly aged (30–40 KYA) samples of an anatomically modern *Homo sapiens*.

One of the major problems with the extraction of DNA from ancient, anatomically modern humans is that because they belong to our own species, it is even more difficult to separate authentic aDNA from the inevitable modern DNA contamination. In these cases, a sequence can only really be considered a valid ancient sequence if it has never been found in any living human population, which may mean that researchers bias their results and potentially throw out target sequences. In an attempt to circumvent this problem, Serre et al. (2004) tried to amplify Neanderthal mtDNA from the well-preserved remains of four Neanderthals and five early modern humans. They found that each of the Neanderthals yielded sequences similar to previously established Neanderthal mtDNA sequences, but that none of the early modern humans did. Later, Krause and Briggs et al. (2010) used NGS to analyze fragmentation patterns and base misincorporations, to identify genuine aDNA, and to distinguish it from contamination. The approach allowed them to recover a complete mitochondrial genome of a 30 KYA, anatomically modern human from the Kostenki 14 site in Russia. The sequence, while unique, exhibits the mutations characteristic of the mitochondrial haplogroup U2. This makes phylogenetic sense as U2 is found in southern Asia, northern Africa, and Europe, and it is thought to be one of the oldest mtDNA lineages in Europe.

One of the other issues often raised in discussions regarding the Neanderthal-modern human relationship is that of interbreeding. Some have suggested that the remains of a child found at the Abrigo do Lagar Velho, Portugal, may represent a morphological hybrid between Neanderthals and *Homo sapiens sapiens* (Duarte et al. 1999; but see Tattersall and Schwartz 1999). Until recently, genetic evidence was limited to mtDNA, which would not provide any indication of admixture if matings involved only Neanderthal males and modern human females. In 2006, a paper was published reporting the sequencing of over one million base pairs of nDNA obtained from the Vindija Neanderthal remains, which date to 42 KYA (Green et al. 2006). The paper also announced the plans of researchers from the Max Planck Institute for Anthropological Genetics, together with a major biotechnology company, to engage in a project in which they would sequence the entire Neanderthal genome. In May, 2010, a draft sequence of the Neanderthal genome representing four billion nucleotides from three individuals was produced (Green et al. 2010). This composite genome was compared with the complete genomes of five modern (extant) humans representing individuals from different geographical populations. The results suggested that there was evidence of limited Neanderthal admixture with modern humans—with the Neanderthal contribution to the modern human genome being in the range of 2.5 +/– 0.6% (Reich et al. 2010). This result contradicted the earlier mitochondrial data and could indicate that the Neanderthal contribution to the modern human gene pool came exclusively from Neanderthal males, or that due to lineage extinction the Neanderthal mitochondrial lineages that had been part of the human gene pool were lost due to stochastic processes.

In addition, their studies indicated that the admixture between Neanderthals was seen only in non-African populations and was equally present among all non-African groups. This observation suggests that the admixture with modern humans occurred after the major migration of anatomically modern humans out of Africa but before those migrating groups diverged from each other. Although the authors were cautious in their interpretations and accepted the possibility of alternative explanations, such as an undocumented ancient substructure within Africa or that more ancient signals of regional gene flow have been obscured by expanding Neolithic populations, a degree of gene flow between Neanderthals and modern humans seems likely to have occurred outside of Africa.

To date, the Neanderthal Genome Project has resulted in evidence regarding skin, hair color (Lalueza-Fox et al. 2007), ABO blood group status (Lalueza-Fox et al. 2008), lactase functionality (Lari et al. 2010), and bitter taste (PTC) perception (Lalueza-Fox et al. 2009). So, we know that at least some Neanderthals probably had light skin pigmentation, possibly even red hair, that they had blood group O, that they were unable to breakdown lactose, and that they were PTC-tasters. Therefore, they would have been unlikely to have liked broccoli and spinach (PTC-tasters find cruciferous vegetables to be bitter tasting—or more so than do non-PTC-tasters)! These studies of the functional genes in Neanderthals and anatomically modern human remains will no doubt continue and should provide some very valuable data regarding evolutionary adaptations in species of *Homo* that are not accessible any other way. This does, of course, only apply when we actually understand the true functionality of the genes in question. For example, we know that the Neanderthal FOXP2 gene is identical to that found in modern humans. This does not, however, tell us about Neanderthal language or communication abilities. The modern human gene sequence is necessary, but is not in and of itself sufficient, for the production of speech and full communication observed in normally functioning, modern humans (Enard et al. 2002).

The Denisovans—Another New Hominin?

In addition to the confirmed mitochondrial and nDNA sequences from anatomically modern humans and Neanderthals, a complete mitochondrial genome was recently recovered from a hominin phalanx excavated at Denisova Cave in the Altai Mountains of southern Siberia (Krause and Fu et al. 2010). The hominin species involved is uncertain, and the phalanx could be older than 50,000 years or as young as 23,000 years. Both Upper and Middle Paleolithic artifacts occurred in the same deposits.

The obtained mtDNA sequence was aligned to those of fifty-four modern humans, the late Pleistocene Kostenki sequence, six complete Neanderthal mtDNA sequences, one bonobo, and one common chimpanzee. The Denisovan differs from that of modern humans by almost twice as many mutations as that of Neanderthals. A phylogenetic analysis suggests that the ancestors of the Denisovan individual diverged from those of Neanderthals and modern humans approximately one million years ago. The authors point out that this individual lived at the same time that Neanderthals were occupying areas to the west, which suggests that at least three distinct hominin lineages were present in Eurasia until at least 40 KYA. This work is the first time a potentially new hominin species has been identified based on DNA alone, and further work and dating at the Denisova Cave site will help us to better understand the chronology.

By the end of 2010, the research group studying the Denisova Cave remains had produced a complete genome (at 1.9x coverage) from the finger bone and another mitochondrial genome from a molar recovered from layer 11.1, just below that from which the phalanx was recovered (Reich et al. 2010). The mtDNA from the tooth was different from that obtained from the finger bone at two positions indicating that it belonged to a second individual, most likely of the same population as the first specimen. It was estimated that the two samples were separated from a common ancestor by approximately 7,500 years. Reich and his team dubbed this population *Denisovans* and suggested that it might have been relatively widespread in Asia during the late Pleistocene. The complete nuclear genome sequence obtained from the Denisovan finger bone indicates that, contrary to the findings based on data from the mtDNA genome, the Denisovans were a sister group to the Neanderthals. Based on a human-chimpanzee divergence of 6.5 MYA, it was estimated that the Neanderthals and Denisovans diverged around 650 KYA, and these two groups shared a common ancestor with modern humans around 804 KYA. It appears that once they diverged, there was limited-to-no contact between Denisovan and Neanderthal populations. But given that we only have two samples from a single site, this interpretation may change in the future as more samples become available and are analyzed.

The Denisovan tooth, which is an upper-left, second or third molar, is very large. Morphological analyses indicate that it is outside of the range of later *Homo* species. If it is a third molar, then it is morphologically comparable to Australopithecine dental remains. If it is a second molar, it is most similar to those of *Homo habilis* or *Homo erectus*. Several morphological characteristics also distinguish it from Neanderthal molars. Reich and others further argued that these primitive traits indicate that the Denisovans split from the Neanderthal lineage at some point before about 300,000 BP when the typical, western Eurasian Neanderthal dental characteristics emerged. This is a time that is consistent with the molecular dates of divergence.

Neither the tooth nor the phalanx have been directly dated because it is said that they would provide too little organic material for both DNA and dating. Reich's research team did report that seven other animal bone fragments found in close proximity to the two samples have been directly dated. These indicate that layer 11 of the cave most likely represents two different occupation periods; one that dates to 50,000 BP or older and another that dates to between 23,000 and 30,000 radiocarbon years BP. The authors suggest that the two hominin samples date to the earlier period.

Perhaps the most surprising result provided by the studies of the complete nuclear genomes of both the Denisovans and the Neanderthals is the evidence for admixture with modern human populations. Where mtDNA studies of both groups indicated that the mitochondrial genomes fell well outside the variation seen in modern humans and that the populations diverged from modern humans hundreds of thousands of years ago, both the Neanderthal and the Denisovan nuclear data indicate that some admixture with *Homo sapiens sapiens* did occur. As discussed above, the Neanderthal findings suggest an approximate contribution of 2.5% Neanderthal DNA to Eurasian populations. The Denisovan data indicate that they did not contribute genes to present-day Eurasians; however, Reich and colleagues do suggest that approximately 4.8% of the genome of modern-day, Near Oceanic peoples from New Guinea and Bougainville can be traced back to Denisovans. They suggest a scenario where at some point after Neanderthals and Denisovans diverged, admixture occurred between Neanderthals and the ancestors of all non-African populations. This was followed by later admixture between Denisovan populations and the ancestors of some Melanesian populations (Figure 5.1).

The most recent study by Reich et al. (2011), which involves analyses of an additional thirty-three modern human populations from Southeast Asia and Oceania, found evidence of Denisovan admixture in the genomes of several other individuals from Island Southeast Asia (ISEA) and the Pacific, including in Polynesia and the Mamanwa, a *Negrito* group from the Philippines. They did not, however, find admixture in mainland East Asian populations, western Indonesians, or in the Jehai or Onge, Negrito groups from Malaysia and the Andaman Islands. Further sampling and, ideally, further aDNA research in Asia and ISEA, will perhaps allow for the identification of when and where that admixture was likely to have occurred.

These results are provocative. Ancient DNA studies from the last few years have challenged our understanding of human evolution in ways similar to the challenges provided by Sarich and Wilson in the 1960s.

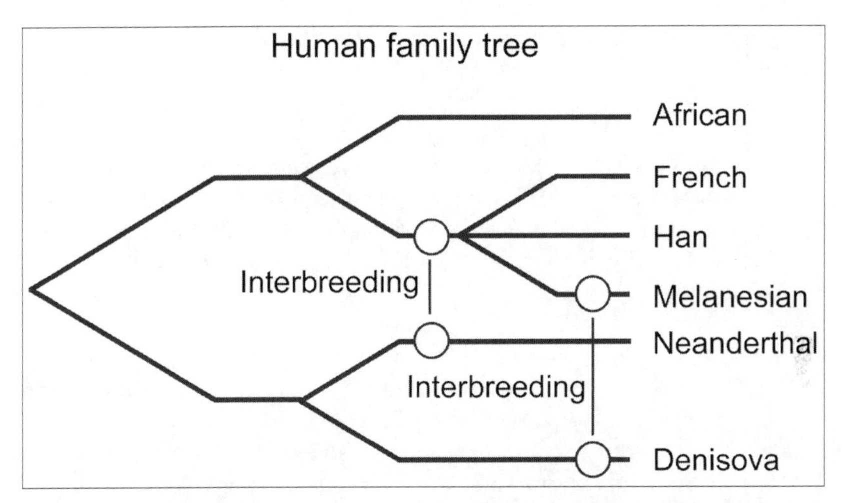

Figure 5.1 A phylogeny showing likely admixture of the non-African branch of modern human populations and Neanderthals, thought to have occurred approximately 80,000 BP and between the ancestors of modern "Melanesians" and Denisovans, which is thought to have occurred approximately 40,000 BP (after Reich et al. 2010).

As the results from Denisova Cave suggest, we might be significantly underestimating the number of hominin species present at any point in time, or alternatively, we may have to reconsider the taxonomic designation of several groups, including Neanderthals. While the potential for aDNA studies to address hominin histories is substantial, genetic data are best interpreted in the context of morphological and archaeological data. This point highlights the need for specialists in a wide range of fields to work together. It also requires that field researchers in paleoanthropological research consider the fact that DNA analyses may also be able to contribute to their research agendas and that they consider the possibility of collecting samples for aDNA analyses under the conditions described in chapter three. This is particularly the case given the rapidity with which technology is developing and opening up previously unimagined research opportunities.

The Hobbit—*Homo floresiensis*

Excavations on the island of Flores in the Indonesian archipelago have revealed a stratified sequence of hominin, faunal, and lithic remains from the cave site of Liang Bua (Brown et al. 2004; Morwood et al. 2004). A suite of dating techniques have indicated that the cave was occupied between 95 and 17 KYA (Morwood et al. 2009; Westaway et al. 2007) by a small-bodied hominin identified, on the basis of a mosaic of morphological features not shared with any other known hominins, as a new species, dubbed *Homo floresiensis* (Brown et al. 2004; Martinez and Hamsici 2008). Particularly striking are the morphological characters of a partial adult skeleton (LB1) with a small brain size (380–410 cm^3), short stature (1 m), *Homo erectus*-like facial architecture, and australopith-like features of the postcranial anatomy (Jungers et al. 2009; Larson and Jungers et al. 2007; Tocheri et al. 2007). Flores may have been occupied by hominins since about 880 KYA (Brumm et al. 2006; Morwood et al. 1998; van den Bergh et al. 2009), and the identity of those first hominins on the island is generally assumed to be *Homo erectus*.

DNA has been extracted from faunal remains—pigs (Larson and Cucchi et al. 2007) and rodents (Matisoo-Smith, unpublished data), though not from the hominin-bearing layers. To date, attempts to extract DNA from the Flores hominins have been unsuccessful (Morwood, personal communication, 2011); however, excavations are ongoing and the team remains hopeful that additional samples with better organic preservation will be recovered.

Mungo Man and Ancient Australians

The fact that DNA can be successfully extracted from hominins as old as 60–70 KYA (Briggs et al. 2009) suggests that we might be able to apply these new aDNA techniques to the study of other early *Homo sapien* specimens to address outstanding issues of human dispersal, such as the initial settlement of Sahul (Australia, New Guinea, and Tasmania). One goal is to better understand the morphological variation demonstrated in the early Australian human remains. This question was addressed initially by Adcock et al. (2001) when they reported the successful extraction and amplification of DNA from ten out of twelve specimens from the sites of Lake Mungo and Kow Swamp in southern New South Wales and northern Victoria. The specimens included the oldest known Australian human remains. Among them was skeleton LM3, which was thought to be as old as 60 KYA but is known date to about 40–45 KYA (Bowler et al. 2003). They generated 354 bp of the hypervariable region of the mtDNA sequence from each of the ten specimens and compared the sequence results with sequences from living aboriginal Australians, the Mezmaiskaya, the Feldhofer Neanderthals, the CRS, and twenty-one sequences of modern Africans representing all known mtDNA lineages. The ancient sequences differed from the CRS by between 2 and 10 bp. For the most part, their analyses did not identify any genetic separation between the early gracile and later robust Australian specimens, which the authors suggested meant that the morphological variation seen in ancient Australian remains must have developed after the date of the MRCA of Australians. Most importantly though, they did report that the sequences recovered from LM3 fell not only outside the range of modern aboriginal Australian mtDNA variation but outside of all present-day human mtDNA variation.

These results created significant controversy (Colgan 2001; Cooper et al. 2001; Groves 2001; Trueman 2001) as the authors claimed that "anatomically modern humans were present in Australia before the complete fixation of the mtDNA lineage now found in all living people" (Adcock et al. 2001:537). Much of the controversy focused on the antiquity of Mungo 3, which had been dated to 62 KYA (Thorne et al. 1999) and was, at the time, significantly older than anything from which DNA had been reliably recovered. The authenticity of the recovered DNA was questioned both because of its great antiquity and because Thorne and his colleagues found that the DNA recovered from LM3 "belonged to a lineage that only survives as a segment inserted into chromosome 11 of the nuclear genome, which is now widespread among human populations" (Adcock et al. 2001:537). Insertion is a natural process as sometimes bits of DNA move about the genome and insert elsewhere. These are called NUMTS or Nuclear inserts into the mt genome. Thus, the sequence produced from LM3 was not identical to known variation in the chromosome 11 nuclear insert, but the phylogenetic analyses of the sequences clustered it with this insert. Perhaps the biggest objection and concern about this reported result is that it has never been independently replicated in another laboratory. Until that happens, these results will remain questionable. It has been argued that, among other problems, the thermal history of the Kow Swamp and Lake Mungo sites make recovery of aDNA, particularly using the methods available in 2001, highly unlikely (Smith et al. 2003). Some of the remains from these sites were repatriated to their descendent communities in southeastern Australia who now maintain control of them. Funding has recently been obtained to undertake new aDNA studies of Mungo Man and other ancient Australians from Willandra Lakes. It is hoped that with the application of NGS techniques that the debates about Mungo Man and the range of morphological variation in ancient Australians will be settled. The recent whole-genome sequence obtained from a 100-year-old Australian Aboriginal hair sample (discussed in chapter eight) has already provided some data to address the issue of continuity of the occupation of Sahul, as well as the relationships between the ancestral populations of Australia and those of other regions.

Chapter Six

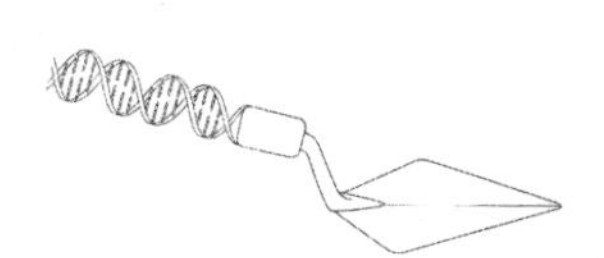

Population Origins and Dispersals

Over the past thirty years, a significant amount of data on mtDNA and Y chromosome diversity from many populations across the globe has been generated and analyzed. Scientists from various disciplines are now able to utilize these data and provide relatively robust interpretations on the origin of these groups and their migration patterns. As discussed earlier, initial studies focused primarily on mtDNA variation, and when Y chromosome data became available for similar populations, it was recognized that, in many populations the migration patterns for women were distinctly different from the migration patterns of men. Most recently, whole-genome analyses, undertaken through the use of SNP chips, have also been used to investigate population origins and migrations. However, whole-genome data are only available for a limited number of populations, and their interpretations are complicated by the recombination events that chromosomes undergo. As more data from diverse populations are collected, whole-genome data will become more useful and will significantly add to our understanding of human migration patterns. Despite the value of these autosomal studies, the problem of recombination will mean that both mtDNA and Y chromosome analyses will still be needed to provide key information about sex-specific dispersal behavior and other aspects of population history that cannot be tracked with the same precision using nDNA markers.

An alternative approach to tracking human migrations is studying genetic variations in a range of organisms that were transported by or comigrated with humans as they moved across the globe. Phylogenies of these commensal species can be used as a proxy for tracking human mobility and may provide insights not apparent when investigating the human genetic data alone.

DNA for Archaeologists by Elizabeth Matisoo-Smith and K. Ann Horsburgh, 109–138.

Some of the most exciting opportunities lie in studies of aDNA, which bring genetic and archaeological data directly together. Such studies have over the last few years produced results that have required revision of many of our explanations of events that have led to the patterns of genetic variation we see in populations today.

In this chapter, we present a brief overview of the current genetic evidence obtained from the analyses of mtDNA, Y chromosome, and other modern nuclear data for humans. We also present information about a range of companion species, as well as how aDNA is used for tracking population origins and migrations around the world. These combined genetic data can be used to test hypotheses and models of population origins and migration patterns generated from the archaeological record or from linguistic data.

"Recently Out of Africa"—The Origins of Modern Humans

As was briefly discussed in chapter one, the first worldwide survey of human mtDNA variation was undertaken and the results published in 1987 by Cann, Stoneking, and Wilson in their paper "Mitochondrial DNA and Human Evolution." This work was done before the development of PCR, so large volumes of tissue were needed to obtain enough DNA for analysis. Whole-genomic DNA was extracted from placental tissue and spun through a cesium chloride gradient to isolate the mitochondria for analysis. While direct DNA sequencing was technically possible in the 1980s, it remained slow, expensive, and tedious and was, therefore, not feasible for the large number of samples with which Cann and colleagues were working. Instead, Cann et al. (1987) digested the extracted mtDNA with enzymes. The enzymes cut DNA when they encounter specific DNA sequences. This technique, known as RFLP analysis (see discussion in chapter two, particularly Box 2.2), is low-resolution because it reveals the presence or absence of a known, particular four-to-six-base, recognition site for a given restriction enzyme. It does not, however, reveal anything about the variation in the DNA sequence in between the restriction sites (where the enzymes cut the DNA).

By comparing the numbers and sizes of the DNA fragments into which the 16.5 kilobases of the mitochondrial genome was cut, Cann et al. (1987) were able to reconstruct the phylogenetic relationships between the 147 women who had donated their placentas. As discussed in chapter two, in order to determine the ancestral relationships within a phylogenetic tree, one must identify the root, or the common ancestor for all members of the tree. Because one of the two deepest branches in their phylogeny led to an African-only clade and because African samples were also distributed throughout the rest of the tree, Cann et al. (1987) decided to root the tree at the mid-point between the two main branches. They argued that the mid-point rooting of the tree also meant that the number of necessary intercontinental migrations was minimal. The results of their analysis suggested that the common ancestor of all the tested mtDNA was African, and that this common ancestral sequence dated to about 200,000 years ago. This ancestral sequence was soon attributed to an individual popularly dubbed as Mitochondrial Eve. It is important to appreciate that the result Cann and colleagues presented does not imply that there was only a single woman living in Africa at the time, just that the mtDNA of all the modern women tested could be traced back to a single common ancestral mtDNA sequence that must have existed in African populations living about 200,000 years ago. There were undoubtedly many other women, carrying many other mtDNA lineages around at that time, but those lineages became extinct at some point between then and now.

This work was taken by many as the final resolution of the out-of-Africa vs. multiregional evolution debate regarding the origins of modern human populations that was being discussed within the broader anthropological literature (Cann et al. 1987; Eckhardt et al. 1993; Stringer and Gamble 1993; Thorne and Wolpoff 1992; Wolpoff 1996; Wolpoff et al. 1988; Wolpoff et al. 1994; Wolpoff et al. 2000). If modern human origins were traced to Africa and were associated with a date of 200,000 years ago, then anatomically modern humans must have replaced all other hominin species (and/or their descendants) who had been part of the initial out-of-Africa migration associated with *Homo ergaster/erectus* 1.8 MYA. Cann et al. (1987), however, drew criticism not only from many in the general public who had difficulty accepting an African origin for all human populations, but also from analytical specialists who chastised the authors for methodological issues, such as not using an outgroup to root the tree (Darlu and Tassy 1987). However, outgroup-rooting was not possible because there were no comparable data of the same resolution for any of our closest relatives, the great apes. They were also criticized when other researchers subsequently found shorter or more parsimonious trees that did not locate the deepest root in Africa (Maddison 1991; Maddison et al. 1992).

Over the next few years, the methods for phylogenetic analyses and characterization of DNA sequences were further developed. In 1991, another of Allan Wilson's students, Linda Vigilant, and her colleagues analyzed 1,122 bp of DNA sequences from the hypervariable region of the mitochondrial genomes of 189 people (Vigilant et al. 1991). They identified a total of 135 unique mitochondrial types, constructed a parsimony tree from them, and used a chimpanzee mtDNA sequence as an outgroup. Their results supported the original interpretation of an African homeland for all modern human mtDNA. The use of a chimpanzee's mtDNA for the outgroup also allowed a better calibration of the rate of mtDNA evolution and gave a date for the common ancestor in the range of 166–249 KYA. Nevertheless, Vigilant and colleagues were also criticized on methodological grounds (Templeton 1991); although, all subsequent analyses, including those of complete mitochondrial genomes (all 16.5 kilobases), have provided further support for a recent modern human migration out of Africa (Atkinson et al. 2008; Horai et al. 1995; Ingman et al. 2000; Ruvolo et al. 1993). Based on analyses of complete mtDNA genomes, which allow for even further precision in calculating the timing of the most recent common ancestor (TMRCA) of all modern humans, Ingman et al. (2000) estimated the date of 171,500 ± 50,000 BP. This result is also consistent with the paleontological evidence for the presence of *Homo sapiens idaltu*, the immediate ancestor of anatomically modern humans (*Homo sapiens sapiens*), which was found in eastern Africa and radiometrically dated to between 160,000 and 154,000 BP (White et al. 2003). Ingman et al. (2000) also reported estimations for the TMRCA of the clade that contains both Africans and non-Africans, thus providing the minimum date for the migration of people out of Africa, to be 52,000 ± 27,500 BP. By examining mismatch distribution data (see chapter two), they further estimated that European populations underwent a recent population expansion approximately 1,925 generations ago. Assuming a twenty-year generation time, they calculated that this equates to about 38,500 years ago. Most archaeologists would recognize this date as one just predating the disappearance of Neanderthals and one that is concordant with some views regarding the earliest evidence for modern human behavior in the archaeological record (Klein 2009); though debates continue regarding the possibility of earlier evidence for modern behavior dating to 70 KYA or more (Brown et al. 2009; Marean et al. 2007).

In the nearly thirty-five years since Cann et al. (1987) published the first genetic data supporting the recently out-of-Africa hypothesis, hundreds of analyses of mtDNA, Y chromosome, and nuclear markers have largely continued to support the model. The extent to which the molecular studies, in general, have incorporated other anthropological information, and the extent to which this information has been incorporated in archaeological models, is still highly variable. However, in a recent introduction to a volume dedicated to reviews of human dispersals, Renfrew (2010) identified the out-of-Africa hypothesis as an archaeo-genetic success story in reference to the synthesis of genetic, archaeological, and linguistic data in the pursuit of more accurate and nuanced reconstructions of prehistory. Currently, improved bioinformatics and Bayesian techniques, which incorporate a priori information, are allowing evermore specific, model-testing regarding modern human migration events (Atkinson et al. 2009; Endicott and Ho 2008). These efforts will no doubt continue to increase the value of genetic data in addressing specific anthropological questions in Africa, as well as in other geographic regions.

Genetic Variation and Population Migrations within Africa

As discussed above, Cann et al. (1987)—in their pioneering study—suggested that all human mitochondrial lineages trace back to Africa and that the ancestors of all non-African populations today would have left Africa no earlier than 200,000 years ago. Of course, many populations remained in Africa and continued to diversify, adapt to ecological changes, interact, and move around the landscape. Studies on a range of genetic markers indicate that private, or uniquely shared, alleles are found at the highest frequency in African populations. They also show that African, or recently African-derived populations (e.g., African Americans), show the highest levels of within-population variation (Henn et al. 2011; Tishkoff et al. 2009). The patterns of diversity within Africa do, however, provide some very interesting evidence relating to specific regional and population histories (Batini et al. 2007; Batini and Lopes et al. 2011; Quintana-Murci et al. 2008; Verdu et al. 2009).

As Cann and colleagues initially showed, the deepest branches in the human mtDNA tree are found in sub-Saharan African populations, and these lineages have since been designated as belonging to macrohaplogroup L (Chen et al. 2000). To date, eight L haplogroups (referred to as L0–L7; see Figure 2.5) have been identified. Each of the haplogroups contain numerous subhaplogroups, which tend to exhibit complex geographical patterning (Gonder et al. 2007). Eastern and southeastern African populations exhibit the most divergent and, therefore, some of the most ancient mitochondrial lineages. These lineages belong to the L0 branch. Further, these extremely deep lineages are absent in northern, western, and some southern African populations. Khoisan populations in southern Africa, or those groups who speak ancient click languages, exhibit the most divergent and the most ancient mtDNA lineages (Behar et al. 2008). Some of these ancient lineages are shared with eastern African populations, such as the L0d branch, which is seen in both the southern African Khoisan and the eastern African click-speaking groups. It has been suggested that the ancestral homeland of Khoisan people may be in eastern Africa, and that some of the ancient lineages seen in Khoisan culture today are remnants of lineages that were lost among the Sandawe and Hadza populations in eastern Africa (Gonder et al. 2007). Eastern Africans, Tanzanians in particular, exhibit extreme diversity of mtDNA types and possess a number of rare, if not exclusive, ancient lineages such as L0f. This observation has been interpreted as suggesting that eastern Africa was significant in both the origin and diversification of modern human populations (Gonder et al. 2007; Salas et al. 2002).

The general scenario for the diversification of mtDNA lineages in Africa is that people carrying lineages belonging to L0 and its sister branch L1'7, which is the ancestral branch to all other L lineages, first spread south out of eastern Africa. This was followed by groups moving westward, at which point populations began to diversify (Figure 6.1). The only major evidence of the introduction of new non-L haplotypes are found in North and northeastern Africa, where the presence of M1 and U6 suggest back-migrations from the Near East during the Upper Paleolithic with H1, H3, and V coming more recently (e.g., early Holocene) across the Strait of Gibraltar from Iberia (Cherni et al. 2009; Ennafaa et al. 2009; Pereira et al. 2010).

Interestingly, it has been argued that there are significant correlations among linguistic, cultural, and genetic affinities in many African populations, notably amongst central African pygmy groups. These hunter-gatherer populations of central Africa have recently been shown to have significant mtDNA structure with evidence of a distinct east-west split. Haplogroup L1c is dominant in the west, where there is also more evidence of admixture with neighboring Bantu-speaking agriculturalists (Batini et al. 2007; Quintana-Murci et al. 2008). Eastern pygmy populations (often referred to as Mbuti), on the other hand, possess predominantly L0a, L2a, and L5 mtDNA lineages (Batini and Lopes et al. 2011; Verdu et al. 2009). Although consensus about the specifics of population definitions within Africa are disputed and muddy, perhaps some of this confusion could be avoided if archaeologists, cultural anthropologists, linguists, and geneticists could come together to agree on basic terminology for the identification of study populations (see Mitchell 2010 for a discussion of the importance of employing accurate ethnolinguistic labels and documentation of the confusion that has arisen as a consequence of multiple, and compounding, failures to do so).

Patterns of diversity in Y chromosomes are not always concordant with those from mtDNA. There are five major lineages of Y chromosomes found in sub-Saharan Africa: A, B, E, J, and R (Cruciani et al. 2002; Tishkoff et al. 2007; Underhill et al. 2001). Haplogroups J and R are largely confined to eastern and central Africa respectively. Haplogroup E is distributed across much of the continent and may be associated with the spread of pastoral and agricultural communities (Henn et al. 2008;

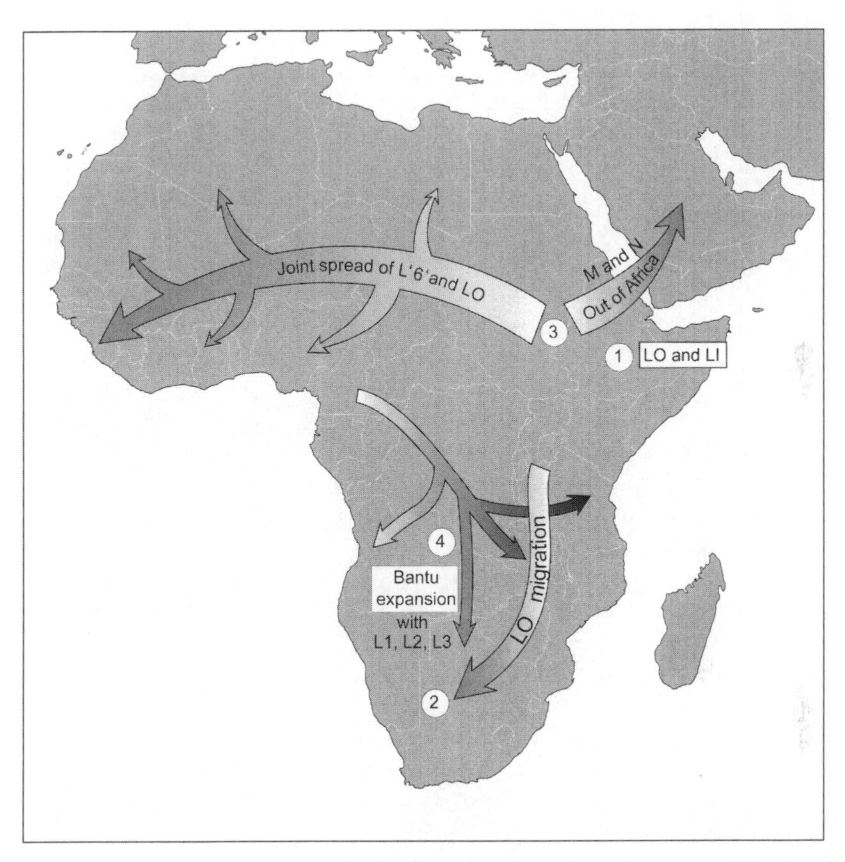

Figure 6.1 Major migration events within Africa based on mtDNA data. (1) The origin of the L0 and L1 lineages in eastern Africa is dated to approximately 150,000 BP. (2) This was followed by the migration of L0 to southern Africa around 140,000 BP, which was followed soon after by (3), the movement of other L lineages into West Africa, and finally (4) the Bantu expansion from West Africa beginning around 3000 BP (modified after Behar et al. 2008).

Underhill et al. 2001). Haplogroups A and B are also widely distributed but are, in contrast, most strongly associated with foraging groups. They are also the most basal, that is the most ancient, lineages of the Y chromosome phylogeny (Batini and Ferri et al. 2011). Recent sequencing of 200 kilobases of the Y chromosome has revealed a slightly different structure of the base of the tree, thereby doubling the number of Y chromosome clades found only in Africa (Cruciani et al. 2011). Additionally, this newly calculated Y chromosome tree suggests a TMRCA of all Y chromosomes to be closer to 140 KYA than the 100 KYA previously estimated (Tang et al. 2002; Thomson et al. 2000; Wilder et al. 2004). In contrast with the mitochondrial data, the oldest lineages in the Y chromosome tree are not found in southern African populations, but in eastern (Batini and Ferri et al. 2011), central (Batini and Ferri et al. 2011; Cruciani et al. 2011), and northwestern (Cruciani et al. 2011) Africa. This suggests that Y chromosome "Adam" lived about 140 KYA and possibly in a different region of Africa than Mitochondrial Eve. These differences in both the timing and location for the human maternal and paternal MRCA may reflect different demographic histories for males and females, but they could also be the result of incomplete geographic coverage in Y chromosome sampling within Africa (Cruciani et al. 2011; Shi et al. 2010).

Investigating patterns of population differentiation and structure in autosomal SNPs distributed across the genome, Henn et al. (2011) found high levels of diversity among Khoisan-speaking Bushmen of the Kalahari Desert. They conclude from these results that modern humans originated in southern Africa. However, when considering this result, one must realize that climatic fluctuations throughout the late Pleistocene induced significant population movements making it likely that the extant inhabitants of the Kalahari are not the descendants of people who have lived continuously in the region for 40,000 years (Hublin and Klein 2011; but see Eren [2012] for an alternative perspective on the effects of late Pleistocene climatic shift on human populations). Additionally, while Henn and colleagues removed the data associated with Bantu-speaking peoples from their analysis, we are nonetheless confronted with the reality that Bantu-speaking agriculturalists expanding out of western Africa some 5,000 years ago (Ehret 2001; Huffman 2009; Mitchell 2002) displaced foraging communities across most of sub-Saharan Africa. It would, therefore, be surprising if the Bushmen of the Kalahari, living in a region unsuitable to agriculture, and thereby less impacted by the Bantu expansion than other groups further north and east, were not the possessors of the oldest lineages. Clearly, this is an area that would greatly benefit from the generation of DNA sequence data from archaeological human remains that predate the Bantu expansion. Unfortunately, many of the relevant archaeological sites provide poor organic preservation, and DNA recovery is seldom possible (Horsburgh, unpublished data from Prolonged Drift and Koobi Fora, Kenya, and Nelson Bay Cave and Klasies River Mouth, South Africa).

While expanding Bantu-speaking populations have obscured signals from deeper time periods, there are considerable genetic data available to complement archaeological and linguistic evidence on the nature of these expansions. Archaeological data indicate that Bantu-speaking populations expanding from southern Cameroon reached eastern Africa by about 500 BCE (Phillipson 2005) and southern Africa by 200 CE (Huffman 2007). The Y chromosome lineages most associated with the Bantu expansion are E1b1a, B2b (Beleza et al. 2005; Berniell-Lee et al. 2009;

Cruciani et al. 2002; de Filippo et al. 2011; Montano et al. 2011), and E1b1a7 (Wood et al. 2005; Zhivotovsky et al. 2004); although, these lineages are found far beyond the current distribution of Bantu-speaking groups today (Rosa et al. 2007). Y chromosome variation across the continent is strongly correlated with variation in language. Interestingly, if the data generated from Bantu-speaking groups are removed from the analysis, the correlation between Y chromosome variation and linguistic variation drops dramatically. This implies that the continent-wide association is largely driven by the patterns associated with the Bantu-speaking groups (Wood et al. 2005).

The mitochondrial lineages associated with the Bantu expansion are L1a, L2a, L3b, and L3e (Salas et al. 2002), and the distribution of these lineages is only weakly correlated with linguistic diversity (Wood et al. 2005). This difference in the pattern between Y chromosome markers and mitochondrial markers among Bantu-speaking peoples suggests sociocultural differences in the spread of males and females. The pattern could have emerged as a consequence of Bantu-speaking men traveling longer distances or in greater numbers than women, Bantu men being more likely to marry local women than vice versa, Bantu groups being highly polygynous (Wood et al. 2005), or perhaps most likely, a combination of all of these.

Recent studies looking at genetic mutations associated with diet are also contributing to studies of pastoralist and agriculturalist mobility and adaptation in Africa and the Middle East. For example, genetic mutations associated with lactase persistence, which allows adults to continue to break down and digest the sugars in milk and other dairy products, have been identified in many pastoralist and dairying populations worldwide (see McCracken 1971 for a discussion of the evolutionary explanation for the distribution). However, recently it has been shown that different mutations occurred in African, European, and Middle Eastern populations, indicating that the mutation occurred in humans more than once (Coelho et al. 2009; Enattah et al. 2008; Gerbault et al. 2009; Tishkoff et al. 2007). Other diet-associated genetic studies such as analyses of salivary amylase production (Perry et al. 2007), which relates to the processing of starchy foods, and PTC-tasting status, which may be associated with avoidance of particular plant toxins (Campbell and Tishkoff 2010; Kim and Drayna 2005), are providing further evidence of early human adaptations and diversity in Africa.

Migrations out of Africa

The MRCA of all non-African mtDNA lineages is clearly traced back to the L3 clade. Recent studies suggest that mitochondrial lineages M and N split from L3 at about 94 KYA (Gonder et al. 2007), and it is estimated that peoples carrying both M and N left Africa at about 65 KYA. It is proposed that the M and N branches split about 63 KYA, and macrohaplogroup R split from N around 60 KYA (Macaulay et al. 2005). Outside of Africa and the Near East, macrohaplogroup M and its derivatives are found exclusively in South and East Asian or Asian-derived populations, including those in Australia and New Guinea. Haplogroup N lineages are found in both Asian and European populations. Two possible migration pathways out of Africa have been proposed based on fossil and mtDNA evidence (Lahr and Foley 1998; Macaulay et al. 2005; Mellars 2006; Watson et al. 1997). One is a northern route through the Nile valley and the Levant, and the other is a more southern route from eastern Africa into the Arabian Peninsula and around toward Asia via the coast. It has been argued that the N and M branches may represent these two distinct expansions out of Africa (Maca-Meyer et al. 2001). In the Y chromosome, the equivalent out-of-Africa lineages are the DE (M1, M145, and M203) and C'F (P143) lineages. However, autosomal data (Melé et al. 2012; Tishkoff et al. 1996; Tishkoff et al. 2009; Xing et al. 2010) and more recent analyses of complete mtDNA genomes (Fernandes et al. 2012) indicate that only a single, initial population migration out of Africa occurred and most likely followed the southern route with diversification occurring within the Arabian Peninsula before populations moved north and east. Later, back-migrations into Africa and population mobility between African, Near Eastern, and European populations added to genetic diversity in those groups (Henn et al. 2012; Moorjani et al. 2011).

Several studies suggest there was a significant population bottleneck leaving Africa (Garrigan et al. 2007; Liu et al. 2006). The most recent estimates of the effective population size for the out-of-Africa migration range from 1,000–1,500 (Garrigan et al. 2007; Liu et al. 2006) to as many as 15,000 people (Zhao et al. 2006). Xing et al. (2010) propose a "delayed expansion hypothesis." This hypothesis purports that once populations left Africa, ancestral Eurasian populations were isolated from African populations for tens of thousands of years, probably in the Near/Middle East, prior to further population expansion. This hypothesis has been supported by recent mtDNA studies (Fernandes et al. 2012), which the researchers suggest is consistent with a demographic refugium or "Gulf oasis" located in the Arabian Peninsula that provided refuge during periods of late Pleistocene hyperaridity (Rose 2010). As discussed in chapter five, studies comparing the complete Neanderthal genome with those of modern humans (Green et al. 2010) indicate that admixture between the Neanderthal and modern human groups occurred after the migration out of Africa, but prior to further non-African migrations, which is consistent with the delayed expansion hypothesis and intermixing in the Near East.

Early Coastal Migrations into and through Southern Asia 40–60 KYA

There is now strong evidence based on mtDNA, Y chromosome, and whole-genome data supporting an early migration along the coast of southern Asia and eastward as far as Australia and New Guinea (Figure 6.2). Approximately 60% of the mtDNA haplotypes in India belong to macrohaplogroup M, and it is slightly higher in tribal populations, who are thought to represent the most ancient populations, than in caste groups (Basu et al. 2003; Maji et al. 2009). Evidence of variation within South Asian populations has shown haplogroup M2 and Y chromosome O2a (M95) lineages have the highest frequencies in Austroasiatic-speaking peoples in India (Basu et al. 2003; Kumar et al. 2007; Kumar et al. 2008). These lineages date to approximately 65 KYA (though there are critiques that these estimates are too early, see Chauby et al. 2011). The distributions of Y chromosome haplogroups C (M130) and D (M174) are similar to the distribution of the haplogroup M mitochondrial lineages, and today they are primarily found in Asia and in East Asian-derived populations. These too are claimed to be likely markers for an early southern migration pathway into and through Asia (Armitage et al. 2011; Stoneking and Delfin 2010; Thangaraj et al. 2006).

Mitochondrial DNA haplogroups derived from the N macrohaplogroup are the predominant lineages in western Asian and European populations. They are also found in India and elsewhere in East and Southeast Asia. Derivatives of haplogroup N, in particular haplogroup P, are also found in Sahul (Australia and New Guinea), which suggests that they were part of the early migration from Africa, despite their currently low frequencies in southern Asia (Macaulay et al. 2005).

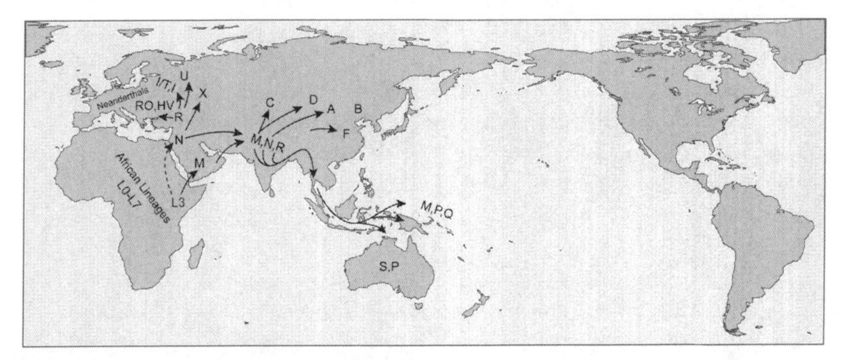

Figure 6.2 Early out-of-Africa migrations and dispersal of mtDNA haplogroups 65 KYA to 30 KYA (modified after Forster 2004).

The fact that Indian populations have a number of unique and ancient mtDNA lineages suggests that there was limited maternal gene flow either in or out of the region after populations arrived during an early out-of-Africa migration event (Metspalu et al. 2004; Thangaraj et al. 2006). Analyses of Y chromosome diversity, however, tell quite a different story (Kumar et al. 2007). It appears that the migration history for Indian males has been influenced by several migration events, including the arrival of Indo-European speakers from central Asia perhaps as recently as 3500 BP (Cordaux et al. 2004). In addition to the markers for early southern migrations (haplogroups C and D), many of the Y chromosomes in India are more closely related to lineages that originate in central Asia and Europe (such as R1a and R2). Thus, they were likely to be more recent arrivals. This is particularly true for members of the caste system in general, as opposed to the tribal groups. There are both north-to-south and caste-rank differences in the Y chromosome diversity and affiliations (Chaubey et al. 2011; Majumder 2010). Populations in the north of India and those from higher castes show closer affiliations with Europeans. Lower castes and southern Indian populations, including most tribal groups, are more closely linked to East Asian populations.

Through analysis of 405 genomic SNP markers, researchers have also reported high genetic diversity with extensive population structure within India (Reich et al. 2009). It has been suggested that this high-level of population structure is due to the deep history of the settlement of South Asia generally and to the antiquity of endogamy between the different religious-, caste-, and linguistically defined populations in the Indian subcontinent (Majumder 2010).

The Settlement of Sahul

Much of ISEA was joined to the mainland during the Pleistocene period, forming the greater continental land mass known as Sunda. This region and neighboring Sahul (the ancient continent formed by the joining of Australia, Tasmania, and New Guinea) (O'Connell and Allen 2004) have been particularly interesting for understanding early human migration out of Africa (Figure 6.3).

Some of the earliest evidence for the presence of modern humans outside of Africa is found in Australia and New Guinea where archaeological evidence of human occupation is dated to at least 45,000 to 50,000 years ago (O'Connell and Allen 2004; Summerhayes et al. 2010). During the Pleistocene period of lowered sea levels, populations were able to follow the coastline of southern Asia on foot until reaching the southern edge of Sunda. From there, the shortest of the crossings between Sunda and Sahul would have required voyages of no less than

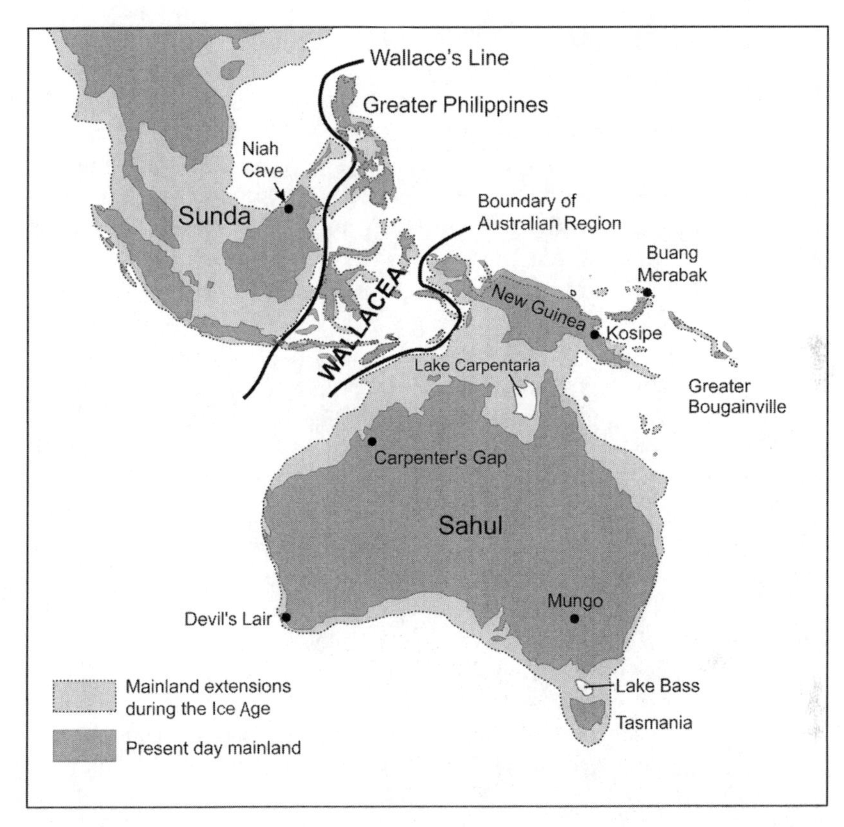

Figure 6.3 Map of Sunda and Sahul during the Pleistocene (dotted lines) and today with some key early (~40,000 BP or older) archaeological sites identified (modified based on O'Connell and Allen 2004).

70 km when the sea level was lowest. Thus, there was likely some form of watercraft used to colonize Australia (O'Connell and Allen 1998). Molecular evidence is consistent with the archaeological evidence of the early arrival of humans in Sahul. Both M- and N- derived lineages are found in the region, and molecular estimates suggest that they arrived there and diversified sometime between 50 and 30 KYA (Friedlaender et al. 2007). Ancient mtDNA from the M and N lineages are also found at low frequencies in isolated populations along the southern coastal route through Asia, as far as Sahul, and in the indigenous communities of the Andaman Islands and elsewhere in ISEA (Macaulay et al. 2005; Mona et al. 2009; Tabbada et al. 2010; Thangaraj et al. 2003).

The regionally specific mtDNA and Y chromosome haplotypes that are found in different parts of Sahul and Near Oceania (which includes the islands of the Bismarck Archipelago through the Solomon Islands chain) suggest that the founding populations were probably relatively small and became isolated not long after arrival. In Australia, the predominant mitochondrial lineages belong to haplogroups S and some unique P haplotypes (both derived from N)—which are not found in New Guinea or Near Oceania—as well as Y chromosomes belonging to ancient C4 (M347) and ancient K (M9). In New Guinea and the islands of Near Oceania, several unique mtDNA lineages from haplogroup M, such as M27, M28, and M29, as well as P and Q haplotypes are found. Yet, a large proportion of the Y chromosomes in the region belong to ancient M1 (P34) and K (M9) lineages (Kayser et al. 2008). Better and more focused sampling in ISEA is beginning to elucidate the dispersal of these early migrations throughout the region where the ancient migration markers are now generally found only at low frequency and in isolated Negrito populations (Delfin et al. 2011; Gunnarsdóttir et al. 2011; Tabbada et al. 2010; Thangaraj et al. 2006).

The archaeological and genetic data indicate that some of the early inhabitants in the Pacific region continued to undertake some major open-ocean voyages. There is evidence of humans reaching the islands of the Bismarck Archipelago by 40 KYA and probably out as far as the end of the main, Solomon Island, chain by 29 KYA (O'Connell and Allen 2004; Wickler and Spriggs 1988). These settlements marked the edge of this early period of human occupation of the Pacific, and populations did not settle beyond the Solomon Islands into the region referred to as Remote Oceania until the last few thousand years.

Colonization of Europe and Western Asia

While the mtDNA, macrohaplogroup M, lineages are exclusively Asian, the haplogroup N and N-derived lineages are found in Asia and Europe. These latter groups appear to be associated with the arrival of Upper Paleolithic technologies and the disappearance of the Neanderthals in western Eurasia. Archaeological evidence suggests that behaviorally and anatomically modern humans arrived in Europe some 45 KYA where they coexisted with Neanderthals—who arrived from an earlier out-of-Africa migration—until about 30 KYA (Higham et al. 2011). Mitochondrial DNA studies indicate that human expansion into the Eurasian continent probably began with the N and R lineages, which dispersed slowly inland from the Near East or western Eurasia, perhaps along major river valleys. This may have been facilitated by climatic changes by about 50 KYA and the retreat of desert conditions along the Levant (Frumkin et al. 2011). These migrations are tracked by the distribution and diversification of haplogroup U, the oldest branches of which are U5 and U8 with basal dates of between 30 and 50 KYA. One of the U branches, U6, later moved back into northern Africa, while U5 and U8 moved into Europe. Other R-derived lineages, such as the J/T and R0 (previously identified as pre-HV) lineages began to diversify in southwestern Asia, and from there they spread into Europe (Pereira et al. 2005; Soares et al. 2010).

Population sizes in Europe seem to have remained relatively small through the late Pleistocene period, though there is some evidence of small population migrations from the east. The changes in climatic conditions with the Last Glacial Maximum (LGM; around 20 KYA) resulted in isolation of populations in small southern refugia around the Mediterranean, the Balkans, and the Levant (Gamble et al. 2004; Soares et al. 2010). The re-expansion into and resettlement of northern Europe and the retreat of the glaciers some 15 KYA left a strong expansion signal in the mitochondrial lineages of modern Europeans. Populations expanding northward from southwestern refugia took with them the newly mutated mitochondrial lineages in the U, H, and V haplogroups (Gamble et al. 2004). Interestingly, the distribution of haplogroup U5b1b1 is shared between the Saami populations in the Arctic and the Berbers of northern Africa, indicating population expansions both north and south from a southern European refugia (Achilli et al. 2005; Tambets et al. 2004). Eastern European populations share most of the same mtDNA haplotypes with Western Europeans; however, some U4 lineages may have come from an eastern European refugium (Malyarchuk et al. 2008).

European Y chromosomes can be traced back to ancient lineages E and F, which diverged from an African source around 70 KYA. Haplogroup E lineages are found in Africa, West Asia, and Europe, whereas F lineages are more widespread outside of Africa. The F lineage diverged into haplogroups I/J, K, H, and G (Cruciani et al. 2007; Rootsi et al. 2004; Semino et al. 2004; Soares et al. 2010). Like the mitochondrial evidence, European Y chromosomes show evidence of diversification during the LGM followed by substantial geographic expansion (Soares et al. 2010). Unlike the mtDNA picture, however, one of the major Y chromosome lineages found in Europeans today, haplogroup R1 (M137), does not appear to be a remnant of one of these expansions from a European or Eurasian refugium. Instead, it may have an origin in southern or southwestern Asia (Kivisild et al. 2003; Soares et al. 2010).

Neolithic Expansion in Europe

While the plants and animals managed by the earliest farmers in Europe are largely of Near Eastern origin, the mtDNA evidence suggests that far from the local hunting and gathering populations being displaced by incoming farming groups, only about 15% of the genetic diversity of modern Europeans can be traced to mtDNA lineages brought by farmers from the Near East (Richards et al. 1996). However, analyses of aDNA from samples across central and northern Europe provide some evidence for genetic discontinuity between hunter-gatherers and early Neolithic farmers in northern-central Europe. Bramanti et al. (2009) found that 82% of the twenty-two hunter-gatherer individuals in their study carried haplotypes belonging to haplogroup U, which are only found at low frequencies in northern and central European populations today. This work further highlights the importance of aDNA research in providing a diachronic perspective on the development of modern European populations. In other words, population history is complex and the people living in a location today are not necessarily the descendants of those who lived there in the past. Population replacement and abandonment of a region due to warfare, climate change, or other natural catastrophes happened in the past and continues to happen today.

Interestingly, the demic diffusion hypothesis has also been suggested by a recent analysis of the most common Y chromosomes found in Europe today, R1b1a2 (R-M269; Balaresque et al. 2010). Prior to 2011, this haplogroup was known as R1b1b2. It was also previously assumed that this haplotype was of ancient Paleolithic origin in Europe, which was an interpretation primarily based on its high frequency in Western Europe. However, Balaresque and colleagues argue for a very rapid and relatively recent (within the last 10,000 years) expansion of the lineage from Central Turkey. The Y chromosome haplogroups E3b (M78), F (M89), and J2 (M172), however, are clearly associated with the Neolithic expansion; although, they are largely limited to southeastern Europe. Finally, N and I lineages appear in Europe from the east, with N3 coming from Siberia, N2e associated with the spread of the Finno-Ugric languages (Rootsi et al. 2007), and I1b* being traced back to the east Adriatic region (Rootsi et al. 2004).

While aDNA studies of Europeans have identified some significant differences in the mtDNA map of Europe at different time periods (Haak et al. 2005; Haak et al. 2010; Melchior et al. 2008), they have much to contribute to the debate regarding the relationship between the development of lactose tolerance and dairying. Lactase persistence is often touted as one of the best examples of gene-culture coevolution (McCracken 1971), but debate has continued around the issue (Beja-Pereira et al. 2003). Was the high frequency of adult lactase persistence in northern European populations the result of natural selection associated with the arrival of dairying practices, or was it the result of dairying being taken up most readily by people who already had the mutation? Ancient DNA analyses from several sites across Europe that date to the Mesolithic and Neolithic, including one Medieval individual for a control (Burger et al. 2007), indicate that lactase persistence was rare among humans. Both the Mesolithic and Neolithic individuals showed an absence of lactase persistence markers, despite the presence of cattle throughout the Neolithic period.

The complexity of European prehistory clearly demonstrates that linguistic, material culture, and genetic attributes do not always align to tell a unified story for the whole of any region. The contribution of DNA studies has also been less than we might hope for in understanding major, ancient migration events in Europe. There have been, however, some valuable lessons learned when very specific questions have been asked. The application of genetic analyses to study more recent events like the Viking expansions into Iceland are providing interesting evidence regarding patterns of specific types of population interaction.

Viking Conquests and Expansions

Several studies in the past decade have assessed the genetic impact of Viking expansions by comparing patterns from both mitochondrial and Y chromosome perspectives (Goodacre et al. 2005; Helgason et al. 2000; Helgason et al. 2001; Helgason et al. 2009; Jorgensen et al. 2004; Wilson et al. 2001). Through their ability to identify typically Scandinavian, rather than typically British/Irish, mtDNA and Y chromosome haplotypes, these studies have indicated quite different settlement histories and Viking influences for populations in both Iceland and the British Isles. Using Norway as an approximation of a typical Viking population, mtDNA lineages belonging to haplogroups I and K and Y chromosomes belonging to R1a and I1d could indicate the influx of Viking populations. The Shetland Islands, for example, show virtually identical proportions of Scandinavian and, therefore, Viking mitochondrial and Y chromosome lineages (about 44%) suggesting relatively balanced sex ratios in colonizing groups. Populations in the Orkneys and from the north and west coastlines of Scotland show lower percentages of Viking markers, but again approximately equal input of male and female Scandinavian markers into the populations (Goodacre et al. 2005). These results indicate that Viking expansion into these areas was likely to be the result of families settling and thus bringing equal numbers of Scandinavian mitochondrial and Y markers to the region. This is in contrast to the studies in Iceland, the Faroe Islands, and the Isle of Skye, which show significantly higher levels of Viking-associated Y chromosomes than Viking-associated mitochondrial lineages (Goodacre et al. 2005; Jorgensen et al. 2004), suggestive of the classic "pillage and conquer" often associated with Viking raids. However, it may also be the case that Iceland was settled by small family groups from the British Isles combined with significant male Viking migration (Helgason et al. 2000; Helgason et al. 2001). It has been argued that this pattern suggests that Viking family units settled in regions close to their political center, while largely male groups interacted with more distant regions. Other similar, historic population interactions that have been addressed by genetic studies include tracking male Phoenician traders around the Mediterranean using Y chromosome markers (Zalloua et al. 2008) and identifying historical, religion-based invasions that happened across Europe and the Near East (Adams et al. 2008; Haber et al. 2011).

East Asian Origins

As discussed previously, modern human populations expanded out of Africa and moved eastward through Asia, then south along the coast into East Asia, and further into what is now ISEA. Populations carrying the mitochondrial haplogroup M and the Y chromosome haplogroups C and D were part of this southern migration. East Asians generally show greater genetic diversity in southern populations than in northern mitochondrial and Y chromosome markers. This greater genetic diversity suggests significant, and multiple, south-to-north migrations within Asia (Jin et al. 2009; Kivisild et al. 2002; Li et al. 2007; Stoneking and Delfin 2010). The distribution of Y chromosomes also generally supports the mtDNA data, though it indicates some north to south movement. North-to-south clines have also been observed in some autosomal SNP studies (Tian et al. 2008). Similarly, the expansion of O3 Y chromosome lineages has been associated with male-biased migration southward from China to the Tibeto-Burman-speaking Han groups of Southeast Asia (Abdulla et al. 2009; Wen and Li et al. 2004; Xue et al. 2005).

Central East Asian populations today are generally dominated by mitochondrial lineages A, B, C, D, G, and F (Kivisild et al. 2002; Wen, Xie, and Gao et al. 2004; Yao et al. 2002; and see Figure 1 in Stoneking and Delfin 2010). Northern Asian populations are generally dominated by mitochondrial haplogroups C and D. Southeast Asian groups are dominated by F and B and, in particular B4, B5, and B6 lineages. They also have some M haplotypes, especially in Tibetan and Thai populations, and significant levels of M7, which is also present in Korea and Japan (Li et al. 2007). The frequency of F1 increases throughout ISEA and is the predominant haplogroup in Java, Borneo, and the Moluccas. Mainland East Asian populations in China, Tibet, and Mongolia, as well as Korean and Japanese populations are highly diverse and appear to have been affected by major population migrations during the Holocene period (Cheng et al. 2008; Hammer et al. 2006; Jin et al. 2009; Xue et al. 2005).

There also appears to be a distinction between Siberian populations and those in the rest of East Asia. In terms of their mtDNA, they have a much higher frequency of Haplogroup C. They also possess haplogroup X2 and Z. The haplogroup X2 originated in the Near East yet represents a recent expansion into Eurasia and up into Siberia at some point since the LGM (Reidla et al. 2003; Volodko et al. 2008). Haplogroup Z is believed to have originated in the region (Tambets et al. 2004). The Y chromosome data indicate that Siberian populations have predominantly N, C, and P (and the derived Q) haplogroups with N (TatC) and C (M86) at the highest frequencies. The N haplotypes are believed to have originated in China from where they moved northward to Siberia, undergoing population bottlenecks and becoming further dispersed into eastern Europe (Rootsi et al. 2007).

The distributions of Y chromosome D (M174), like mtDNA haplogroup M7, are found at very high frequencies in Tibet, extending to Korea, Japan, and Thailand; but they are not found in northern Southeast Asia or ISEA (Jin et al. 2009; Stoneking and Delfin 2010; Wen and Xie et al. 2004). The high diversity we see in China and mainland Southeast Asia is due to substantial Neolithic population migration and mobility associated with the development of rice agriculture (Hammer and Zegura 2002; Hammer et al. 2006; Jin et al. 2009). It is interesting to note, however, that a recent study of mtDNA variation suggests that population expansion began prior to the development of rice agriculture and may have been one of the forces that drove agricultural development rather than the other way around (Zheng et al. 2011).

Militaristic expansions through Asia and further afield also had a significant impact on Y chromosome diversity as is seen elsewhere. One of these in particular, which has had a significant impact on the distribution of Y chromosomes not only in Asia but extending into Europe, was the dispersal of the C3 Y chromosomes. This distribution has been tied to the Mongol invasions, also referred to as the Genghis Khan invasions (Zerjal et al. 2002; Zerjal et al. 2003).

Ancient DNA evidence of significant population movement and replacement in Asia is demonstrated in the results of the study of mtDNA variation in three populations dating between 2,500 years old and modern day. The populations are from Linzi, in Shandong Province, central-east China (Wang et al. 2000). Wang and colleagues reported that the oldest samples showed a closer mitochondrial affinity with populations in Europe and, in particular, Turkish, Icelandic, and Finnish populations, than with groups living in Linzi today who show typical East Asian haplotypes.

Siberia was first settled by people who moved northward out of central and east-central Asia and became established in southern Siberia between

45 and 30 KYA. As is the case in China, aDNA from populations around Lake Baikal indicates discontinuity between ancient (early Neolithic) and modern groups (Mooder et al. 2006). When the mtDNA from two Neolithic cemetery populations near Lake Baikal in southern Siberia were compared with DNA from a Mongolian cemetery and that of modern Siberians, the results showed that a major change in genetic composition of the region had occurred. In the ancient samples at the Lokomotiv cemetery near the southern shore of Lake Baikal, the patterns of mitochondrial lineages vary dramatically over time. The burials from the earlier Siberian cemetery appear to have more mtDNA lineages associated with western Eurasian populations, whereas the later Siberian population has an increase in East Eurasian markers. This later group also appears more similar to the modern Mongolian populations. The research team suggests that the region was abandoned sometime around 6000 BP and recolonized by a different population some 800 years later (Mooder et al. 2006).

Settlement of the Americas

To date, five major mtDNA lineages have been identified among the living indigenous peoples of the Americas. These are lineages A, B, and X—derived from macrohaplogroups N, C, and D, which are branches of macrohaplogroup M. Haplogroups A, B, C, and D have long been known to have originated in Asian populations and were brought across the Bering Strait with the founding populations (Figure 6.4). Haplogroup X was initially thought to be an indicator of prehistoric European contact until it was discovered among Siberian populations, making that a more likely source (Fagundes et al. 2008; Goebel et al. 2008). While as yet unreported among living or prehistoric groups native to the Americas, two archaeological samples dating to around 5,000 years ago from an

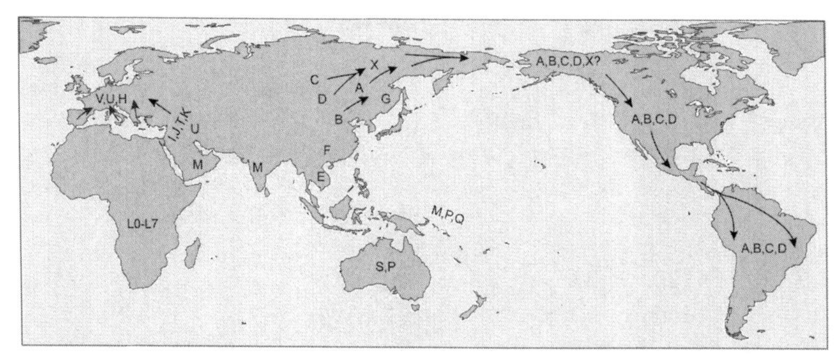

Figure 6.4 Post-LGM (20,000–10,000 BP) population migrations and dispersal of mtDNA lineages (modified after Forster 2004).

archaeological site in British Columbia have been reported as belonging to mitochondrial lineage M from Asia (Malhi et al. 2007). Haplogroup A is found at higher frequency in North America and decreases southward while the reverse is true of haplogroups C and D. Haplogroup B has a patchier distribution and is rare in northern North America and rare to absent in some South American indigenous populations, such as those from Patagonia and Tierra del Fuego. A rare haplotype, D4h3, has been reported in both North and South America, and it appears to have a strictly Pacific coast distribution (Perego et al. 2009).

Regardless of a growing database, there is still some debate surrounding the number of population migrations into the Americas (see discussion in O'Rourke and Raff 2010). On the basis of the mtDNA data from modern indigenous groups, it has been suggested that all five major mitochondrial lineages arrived in the New World as part of a single migratory population moving out of an environmental refugium in Siberia in which they had been isolated between 19 and 15 KYA (Fagundes et al. 2008; Tamm et al. 2007). This is consistent with the current archaeological consensus for a pre-Clovis occupation of the Americas as the earliest dates are now around 14.5 KYA. These dates stem from Monte Verde in southern Chile (Dillehay et al. 2008) and from human coprolites recovered in Oregon, from which human mtDNA revealed founding lineages A2 and B2 (Gilbert et al. 2008). These 14.5-KYA dates are about 1,000 years earlier than the earliest accepted dates for the Clovis complex (Waters and Stafford 2007). It is worth noting, though, that the existence of a single geographic origin for the peopling of the Americas does not necessarily imply that there was only a single migration.

New World, Y chromosome diversity is similarly limited and likewise shows Asian and Siberian origins. Over 80% of modern Native American Y chromosomes belong to lineages Q and C (Blanco-Verea et al. 2010; Santos et al. 1999; Schurr and Sherry 2004; Zegura et al. 2004). While there is some diversity in both lineages, most belong to Q1a3a (M3) and C3b (P39) (O'Rourke and Raff 2010). Not surprisingly, however, modern Native American populations show significant levels of European admixture (Blanco-Verea et al. 2010). Autosomal markers also show limited genetic diversity compared with other populations worldwide but do indicate geographic structure, with declining genetic variability as distance from the Bering Strait increases (Wang et al. 2007).

The Settlement of Remote Oceania and Polynesia

While the Pacific region traditionally has been divided into three geographic zones (Melanesia, Polynesia, and Micronesia), it has been argued that these definitions make little biological, historic, or linguistic sense.

Archaeologist, Roger Green, and linguist, Andrew Pawley, proposed that the Pacific was more logically defined by the regions of Near and Remote Oceania (Green 1991; Pawley and Green 1973). One of the last major human migration events involved the settlement of the islands of Remote Oceania, which began in the last 3,000 years and culminated with the settlement of the extremes of the Polynesian Triangle only 750 years ago.

As discussed previously, humans have been in the Pacific region for at least 50,000 years and as far east as the Solomon Islands for over 20,000 years. However, the earliest archaeological evidence for humans in Remote Oceania does not appear until approximately 3100 BP when people associated with the Lapita cultural complex arrived in the southeast Solomon Islands, Vanuatu, New Caledonia, Fiji, Tonga, and Samoa. Lapita sites are identified in Near Oceania by the appearance of a range of artifacts including the distinctive, dentate-stamped pottery that the complex is best known for along with fish hooks, shell ornaments, and new forms of adzes made of stone and shell. Lapita sites first appear in Near Oceania in the Bismarck Archipelago around 3350 BP and rapidly spread eastward, reaching Tonga and Samoa on the edge of the Polynesian Triangle within a few hundred years (see Figure 6.5). The sudden appearance of Lapita sites in Near Oceania and their rapid spread across the Pacific has been tied to the movement of Austronesian languages. The spread of Lapita sites, in particular, matches the distribution of the Oceanic Austronesian subgroup (Kirch and Green 2001). Eastward expansion of Lapita stopped in west Polynesia for a period of at least 1,500 years, and it was during this time that Polynesian language and society developed (Kirch 2000). The settlement of the Polynesian Triangle itself, defined by the apices of Hawai'i, New Zealand, and Rapa Nui (Easter Island) began only in the last 1,200–1,500 years.

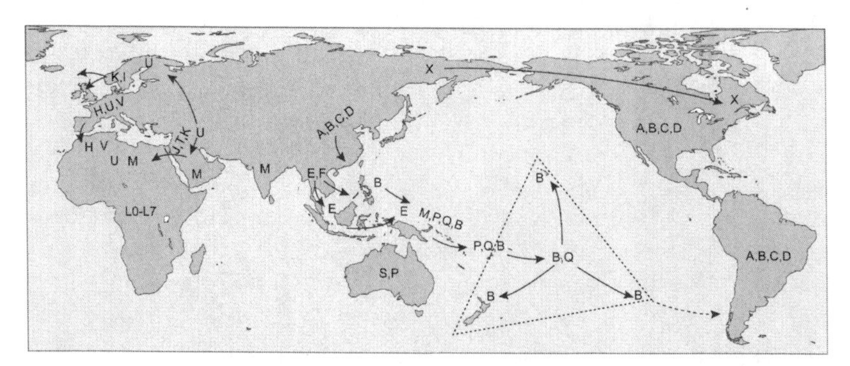

Figure 6.5 Holocene (10,000 BP onward) population migrations and dispersal of mtDNA lineages (modified after Forster 2004) showing migrations to the Polynesian Triangle and possible further contact between Polynesia and South America.

Much research has focused on identifying the origins of the Polynesians, and to date, archaeological (Kirch 2000), linguistic (Gray and Jordan 2000), and genetic data (Gao et al. 1992; Kayser 2010; Serjeantson and Hill 1989) indicate that a large proportion of Polynesian ancestry can ultimately be traced back to Asian or ISEA populations. However, the degree to which admixture with indigenous Near Oceanic populations occurred remains a topic of debate, as does the timing issue of Asian/Oceanic interactions (Friedlaender et al. 2008; Hunley et al. 2008; Kayser et al. 2000; Kayser et al. 2006; Soares et al. 2011). As yet, there are no human, genetic data indicating clear prehistoric contact between Polynesia and South America, though the presence of Native American Y chromosomes have been identified on the island of Rapa in the Austral Islands. These have been identified as historic introductions associated with the 1860s Peruvian slave trade in the Pacific Islands (Hurles et al. 2003). The widespread, prehistoric distribution of the South American sweet potato across Polynesia (Hather and Kirch 1991) and the presence of pre-Columbian Polynesian chickens in South America, however, clearly indicate some degree of contact (Storey et al. 2007; Storey and Quiroz et al. 2008). Given the superior navigational and voyaging skills of the ancient Polynesians and the naturally continued eastward trajectory of Polynesian expansion (Fitzpatrick and Callaghan 2009), this contact was most likely made by Polynesians reaching South America rather than the South Americans voyaging into the Pacific as was proposed and recreated by Thor Heyerdahl (1952).

Early mtDNA studies on Polynesian populations identified extremely high frequencies of haplogroup B4 with particular hypervariable region mutations that were quickly dubbed the Polynesian motif (Melton et al. 1995; Redd et al. 1995; Sykes et al. 1995). Subsequent studies have indicated that this haplotype, now known as B4a1a1a, is not uniquely Polynesian but is found in coastal populations from eastern Indonesia, throughout New Guinea, across the Pacific (Friedlaender et al. 2005; Pierson et al. 2006; Tabbada et al. 2010), and also in Madagascar (Razafindrazaka et al. 2010). While its distribution appears to map the Neolithic expansion of Austronesian populations out of Taiwan (Gray and Jordan 2000; Gray et al. 2009), none of the coalescent dates for the haplotype diversity fit well with the archaeological timing of the appearance of the Lapita cultural complex (Pierson et al. 2006). The most recent analyses suggest that the motif arose in Near Oceania (Soares et al. 2011). In addition to the B haplotypes found in Remote Oceania, the much more ancient P, Q, and M haplogroups are also found in Remote Oceania at decreasing frequencies eastward across the Pacific. Of these Near Oceanic lineages, only Q1 has been found as far east as Polynesia (Deguilloux et al. 2011; Friedlaender et al. 2005; Friedlaender et al. 2007; Friedlaender et al. 2008; Kayser 2010).

While the early mitochondrial data taken primarily from analyses of the hypervariable region alone fit reasonably well with linguistic models suggesting Austronesian expansion out of ISEA, more recent Y chromosome studies (Hurles et al. 1998; Kayser 2010; Kayser et al. 2006; Su et al. 2000) and complete mitochondrial genome sequencing (Pierson et al. 2006) indicate a more complex scenario. Unlike the mitochondrial lineages in Remote Oceania, Y chromosome analyses indicate a substantial Near Oceanic contribution to Polynesian diversity. While the Asian Y lineages belonging to haplogroup O, like O3a (M122), are found throughout Polynesia, higher frequencies of Near Oceanic Y haplogroups C2a (M208), K (P79), and M1 are found in most Remote Oceanic populations (Kayser 2010; Kayser et al. 2006; Kayser et al. 2008; Scheinfeldt et al. 2006). Unfortunately ascertainment bias, due to the fact that most genetic marker screening panels have been designed for European or Asian populations, remains a major problem in studies of both Oceanic mitochondrial and Y chromosome variation, resulting in a current underestimation of the real level of diversity. To date, many of the Y chromosome lineages in the Pacific are not fully resolved (Kayser 2010).

Studies utilizing fine-grained sampling of mitochondrial and Y chromosome diversity within Oceania and ISEA (Friedlaender et al. 2007; Kayser et al. 2008; Mona et al. 2009; Tabbada et al. 2010) are revealing increased genetic diversity within the geographic region, indicating that there may have been significantly more interaction than previously considered. It has also recently been recognized, for example, that interactions with ISEA have resulted in the introduction of mtDNA lineages E1a and E1b to Near Oceania. To date, these lineages have not been identified beyond Near Oceania, but they are markers of relatively recent (Holocene) Asian admixture (Friedlaender et al. 2007). A recent study of autosomal markers in Island Melanesia estimated that Asian ancestry was less than 20% in many of the Austronesian-speaking groups (Friedlaender et al. 2008; Kayser 2010). Thus, once again, the presumed links between language, culture, and biology do not hold up so well in the Pacific, particularly in Near Oceania.

Micronesia has received much less focus in terms of both archaeological and genetic studies, but the settlement history is similar to other parts of Oceania in that both mitochondrial and Y chromosome studies indicate both East Asian and Near Oceanic contributions. In the case of western Micronesia, the earliest contact appears to be from East Asia and dates to approximately 4,000 years ago (Lum and Cann 2000). However, it has been argued that the settlement of much of central and eastern Micronesia (the Caroline and Marshall Islands for example) was associated with post-Lapita population movements from Near Oceania or Vanuatu (Kirch 2000). Yet, recent autosomal studies indicate that Micronesian and Polynesian lineages cluster together. Although separate from Melanesian populations, they are most closely related to East Asian inhabitants, particularly Taiwanese aboriginal groups (Friedlaender et al. 2008). The similarities between Micronesian and Polynesian populations at the genetic and phenotypic level has led some to suggest that their settlement histories are more complex than previously thought. These researchers suspect that there may also be more direct Southeast Asian input into these groups at some point after the Lapita migrations (Addison and Matisoo-Smith 2010).

The disparity between the mitochondrial and Y chromosome histories in the Pacific, it has been argued, may be the result of the matrilineal and matrilocal nature of many Pacific communities (Hage and Marck 2003). However, it is also clear that Pacific populations have undergone significant bottlenecks, not only during colonization but also as a result of post-colonization disasters, including the well-documented historic population crashes associated with the introduction of European diseases. Post-contact admixture with European populations is also dramatically reshaping the variation we see in the Pacific, particularly in terms of the Y chromosome (Hurles et al. 1998; Hurles et al. 2003). Consequently, there is a very good chance that analyses of the genetic variation seen in Pacific populations today may not reflect the level of variation present at the time of initial settlement.

To date, there have been few aDNA studies of Pacific skeletal remains. Early studies in aDNA involved analyses of archaeological remains from Pacific populations, including samples from Lapita contexts (Hagelberg and Clegg 1993; Hagelberg et al. 1994). However, these studies were undertaken before researchers recognized the extreme problems associated with contamination; and because the results have never been repeated, they are generally not accepted as authentic. Recently, however, researchers have undertaken aDNA analyses of Polynesian skeletal remains, which are particularly interesting in their implications regarding ancient mtDNA diversity. Deguilloux et al. (2011) undertook aDNA analyses of seven burials from the atoll of Temoe in the Gambier Islands, French Polynesia. The burials dated from the fourteenth to the seventeenth century and were clearly pre-European. They obtained enough mtDNA sequence data to be able to assign all of the samples to specific mtDNA lineages. While six of the samples belonged to the B4a1a1a type, one of the ancient samples was typed to haplogroup Q1, a Near Oceanic haplotype. Because this lineage is very rare in east Polynesian populations, it is somewhat surprising to find it in a small sample from far-eastern Polynesia. Given that so much variation was identified in only seven ancient samples, this may suggest that the extreme mitochondrial homogeneity claimed for modern Polynesian populations could reflect a loss of diversity due to post-settlement bottleneck events, such as the depopulation many islands endured due to the introduction of European diseases. Only further aDNA studies will tell us how well modern genetic diversity reflects original founding populations in a region even as recently settled as Remote Oceania. Based on previous studies, we can anticipate that such research will likely cause us to dramatically reconsider existing models of settlement.

Proxies for Human Dispersal—Animals, Plants, and Other Organisms

While most human codispersals are associated with humans moving domesticated animals, such as dogs, cattle, sheep, goats, pigs, plants, and other organisms were also transported by or in association with human populations. Studies of these species have been most valuable for tracking prehistoric human migration and interaction, particularly when humans translocated species to regions where they would have been unable to disperse independently, such as the movement of animals to islands. Accordingly, islands provide excellent environments across which to track the movement of plants and animals. Tracing their human-mediated movement across continents is more difficult. The study of the distributions and genetic variation of animals and plants and their use as proxies for tracking human movement has most widely been applied in the Pacific but is increasingly being undertaken elsewhere (Fraser et al. 2012; Hardy et al. 1994; Kasapidis et al. 2005; Larson et al. 2007a; Rick et al. 2009; Suchentrunk et al. 2006). Genetic data relating to translocations in the Pacific will be presented here as a case study.

Translocations in the Pacific—Animals

The earliest documented human translocation of an animal species anywhere in the world substantially predates the earliest evidence for animal domestication and involved the introduction of the northern common cuscus (*Phalanger orientalis*) from New Guinea to New Ireland approximately 20 KYA (Grayson 2001). The translocation of animals to Pacific Islands and the introduction of exotic species dramatically increased in the Holocene period, as it did elsewhere. During this time, several species appear to have been introduced from somewhere in ISEA into the Pacific as part of the Lapita cultural complex. Kirch (2000) describes how transported landscapes were particularly important for Pacific colonists. Through this process, Lapita peoples and their descendants introduced a number of plant and animal species to the islands that they settled. Four animals in particular are associated with the Lapita expansion: dog, pig, chicken, and rat. Both ancient and modern DNA studies of all four of these species have been undertaken in recent years (Larson and Cucchi et al. 2007; Matisoo-Smith and Robins 2004; Savolainen et al. 2004; Storey et al. 2010). While studies of extant populations of commensal animals can be useful in elucidating general patterns, it is essential that aDNA be incorporated in phylogeographic studies of translocated animals, plants, and organisms. This is particularly true in the case of those species that were also transported by later European colonists and traders. Genetic analyses of archaeological faunal remains are a valuable complement to human genetic data. This is because larger sample sizes are often available, and because their analysis is somewhat less constrained by cultural considerations inherent in the destructive analysis of human remains.

Many similarities have been identified in the patterns shown in genetic data from different animal species. For example, mtDNA analyses of rats, dogs, and chickens all indicate multiple introductions into Remote Oceania (Matisoo-Smith and Robins 2004; Matisoo-Smith and Robins 2009; Matisoo-Smith et al. 2009; Savolainen et al. 2004; Storey and Ladefoged et al. 2008). This pattern has not yet been identified in analyses of modern human genetic diversity from the Pacific; although, as discussed previously, aDNA analyses of human remains often reveal patterns not seen in analyses of modern genetic diversity. Data from domesticated pigs, on the other hand, indicate that only a single lineage was introduced into the Pacific (Larson and Cucchi et al. 2007). This finding does not, of course, allow us to distinguish between the possibilities of one or more introductions from the same source population. The analysis of DNA and radiocarbon dating of chicken bones has also identified evidence of Polynesian contact with pre-Columbian South America (Storey et al. 2007; Storey and Quiroz et al. 2008), which supports the previous evidence of contact indicated by the presence of South American sweet potatoes across Polynesia (Hather and Kirch 1991; Yen 1993). Genetic analyses of the Polynesian white-shelled tree snail in the Society Islands have indicated intra-archipelago trading of these snails whose shells were used for personal adornment during prehistory and early historic times (Lee et al. 2007).

Translocations of Plants in the Pacific

Pacific peoples also transported a number of plant species from Asia and Near Oceania. Unfortunately, genetic studies in most cases are limited to extant populations and are, therefore, complicated by recent and historic reintroductions and transport. However, genetic studies have provided some interesting and significant information regarding human choice and cultivation practices for a range of species, including bottlegourd (*Lagenaria siceraria*), ti (*Cordyline fruticosa*), breadfruit (*Ardocarpus altilus*), coconut (*Cocos nucifera*), paper mulberry (*Broussonetia papyrifera*), and sweet potato (*Ipomoea batatas*) (Baudouin and Lebrun 2009; Clarke et al. 2006; Hinkle 2007; Seelenfreund et al. 2010; Zerega et al. 2004; Zhang et al. 2004). It is important to highlight that while some of these were food plants, many were taken for the production of cloth, for the construction of houses and boats, and for use as containers.

While it has been argued that the transportation of *Rattus exulans* into the Pacific was an intentional translocation, rodents have accompanied human expansion elsewhere, and many of these animals have simply been unintentional co-travelers.

In this regard, genetic evidence has been used to trace the movement of mice, voles, shrews, and stoats to the British Isles alongside Iron Age people and Norwegian Vikings (Martínková et al. 2007; Searle et al. 2009). This method has also been used to trace the introduction of *Rattus rattus* to Madagascar from India and the Arabian peninsula (Hingston et al. 2005; Tollenaere et al. 2010). In addition to the plants and animals that people transported, they also carried other organisms. These included pathogens, such as gut bacteria, viruses, and parasites, such as lice, mites, or even malarial protozoa (*Plasmodium falciparum*; Lum et al. 2004; Miranda et al. 2004; Takasaka et al. 2004; Yanagihara et al. 2002). Analyses of genetic variation in these co-travelers are also now being studied in order to identify population origins, interactions, and migration pathways.

Helicobacter pylori and Modern Human Migrations

Colonizing more than half of the stomachs of modern humans around the world, the *Helicobactor pylori* bacterium apparently spread with human hosts during the early migrations out of Africa and throughout the rest of the world. *Helicobactor pylori* is harmless in most people that it infects, but in others it is associated with an increased risk for several types of stomach cancer or peptic ulcers. Once obtained, a person will carry *H. pylori* in their stomach for the rest of life. It has a rapid rate of mutation, making it ideal for tracking the evolutionary history of the various strains. Four major populations of modern *H. pylori* have been recognized worldwide: hpAfrica1, hpAfrica2, hpEastAsia, and hpEurope. The strain, HpEastAsia, is further divided into three subpopulations called hspAmerind, hspEastAsia, and hspMaori. The strain, HpAfrica1, is split into hspWestAfrica and hspSouthAfrica (Falush et al. 2003). Another, HpAfrica2, is largely confined to the Khoisan-speaking, hunter-gatherers of southern Africa, while hspWestAfrica and hspSouthAfrica are both primarily associated with Bantu-speaking farmers. Having studied the genetic diversity of 769 *H. pylori* samples from 51 geographically dispersed ethnic groups, Linz and Schuster (2007) found that the most recent common ancestor of all the *H. pylori* they studied could be traced back to eastern Africa and dated to about 58,000 years ago. In a pattern similar to that of the genetics of its human host, the diversity of the *H. pylori* genome is highest in eastern Africa and decreases with increasing geographic distance from that region. Not surprisingly, complex human behavior is also indicated by a high degree of admixture resulting in horizontal transmission of *H. pylori* in Europe for example with the appearance of strains indicating Neolithic population expansions (Falush et al. 2003). Generally, the European *H. pylori* populations are derived from both northern African and central Asian types. Further studies of *H. pylori* focusing on more specific issues, such as ethnic diversity in India (Wirth et al. 2004) and the history of the Austronesian expansion (Moodley et al. 2009), indicate that *H. pylori* can be a valuable tool for investigating events of shallow time depth. One major disadvantage, however, is that sampling for *H. pylori* is highly invasive and requires a stomach biopsy. Most samples collected so far have been as a result of medical studies investigating stomach cancer or peptic ulcers. If a noninvasive technique for sampling were to be developed, this could be a most valuable tool for tracking population origins and interactions.

JC Virus and Human Migrations

The JC virus (JCV) is a small, DNA virus that is ubiquitous in human populations. It tends to be transmitted vertically, from parents to offspring but has limited horizontal transfer outside families. The JCV strains are classified into eighteen major genotypes with unique geographic distributions. On the basis of complete JCV genome sequencing, three major types have been identified (Hatwell and Sharp 2000; Sugimoto et al. 2002; Yogo et al. 2004). Type A is found in Europe, the Mediterranean, Asia, and arctic North America. Type B is the most geographically diverse type found across Africa, South Asia, Europe, and in Asian-derived populations from the Americas and Oceania. Type C is found only in Africa. The JCV has a mutation rate at least two orders of magnitude higher than human chromosomes. The rapid rate of variability generation makes it particularly valuable for addressing recent events. This property of JCV has been profitably applied in Oceania where, for example, there are regionally specific lineages (i.e., 8a and 8b) and where one (8a) has been found only in Near Oceania (Takasaka et al. 2004).

Conclusion—Broad Patterns

Since the initial development of the field of molecular anthropology, the application of molecular methods has contributed greatly to our understanding of human dispersal out of Africa and around the globe. Data have increased over the years to include more populations, longer sequences, and more markers. Interesting and apparently robust patterns are emerging from multiple markers, including mtDNA, Y chromosome, and autosomal SNPs (Atkinson 2011; Shi et al. 2010). We see, for example, strong geographic patterning at the continental level with the oldest populations, the largest effective population sizes, and the earliest expansion dates all in African populations. We also see the youngest MRCA, most recent population expansion dates, and the smallest effective population sizes in the Americas. Mean STR diversity, MRCA, and effective population sizes were negatively correlated with walking distance from East Africa (specifically Addis Ababa) (Shi et al. 2010).

Analyses of mtDNA, coding region sequences (Atkinson et al. 2008) indicate the following time periods for the initial population expansions of modern humans: 143–193 KYA for sub-Saharan Africa; 52 KYA for South Asia; 49 KYA for the rest of Asia; 42 KYA for Europe; 40 KYA for the Middle East and North Africa; 39 KYA for New Guinea; and 18 KYA for the Americas. Although estimates based on Y chromosomes provide slightly younger temporal designations (on average 76% of mtDNA dates), they result in the same order for the expansion of populations (Shi et al. 2010). While these large patterns are emerging, inevitably local historic events and forces will have had impacts on genetic variation within local populations. Studies of commensal organisms may also be more appropriate for addressing later events (Addison and Matisoo-Smith 2010). It is in teasing apart the more recent events and addressing specific questions that the collaboration between archaeologists and geneticists is, and will be, the most valuable.

CHAPTER SEVEN

Human Impacts—Extinction, Domestication, and Utilization of Plants and Animals

One of the most-well-documented impacts of human expansion across the globe is the local extirpation, or complete extinction, of native flora and fauna. Because humans arrived in the New World at a time that coincided with the extinction of numerous species, including megafauna, researchers have attributed their demise to human impacts. More recent work, though, suggests that the extinction process is more complicated and can be attributed to several causes (Koch and Barnosky 2006; Yule et al. 2009), even possibly extraterrestrial catastrophes (Firestone et al. 2007). This is not to say, however, that the direct effects of the arrival of humans and commensal predatory species have not resulted in rapid extinction in various times and places. This is well-documented on many Pacific Islands (Steadman et al. 2002). Bunce et al. (2009), for example, demonstrated that in New Zealand a combination of factors, including hunting, habitat destruction, and predation by other introduced species (dogs and rats), led to the rapid extinction of the endemic moa. In this regard, aDNA analyses of radiometrically dated bones (Shapiro et al. 2004), and even aDNA recovered from frozen sediments (Haile et al. 2009), can provide key insight regarding the genetic history of species, mapping the decline in population size and diversity in real time. These data can then be compared with the archaeological and environmental data to provide a more specific picture of the timing and potential causes of the extinction events.

The apparent impact of human arrival on megafauna began around 50,000 years ago in Australia, where some sixty large terrestrial species

DNA for Archaeologists by Elizabeth Matisoo-Smith and K. Ann Horsburgh, 139–154.

became extinct (Miller et al. 2005). A similar picture has been painted for the advent of humans in the New World, with evidence suggesting a late arrival. The argument for the role of direct human impact is generally simple. It has typically been thought that rapid overkill of native species, which had never before encountered major predators, resulted in mass extinctions within a thousand years of human occupation (Burney and Flannery 2005; Martin 1984). It has been argued, however, that the timing of the extinction events in the Americas coincided not only with human arrival, but with significant changes in climate. These climate changes occurred at the Pleistocene-Holocene boundary, suggesting that climate rather than humans could have been the real cause of the mass extinctions. Thus, a debate that arose in the 1800s (Grayson 1984) regarding whether it was direct human impact or climate change that caused the demise of megafauna worldwide began again (Koch and Barnosky 2006).

While the debate has been addressed by researchers in a range of fields, including archaeology (Grayson and Metlzer 2002), aDNA evidence suggests that direct human impact cannot be seen as the main cause of megafaunal extinction. In a study of Pleistocene bison remains from archaeological sites across China, Siberia, Canada, and the United States, Shapiro et al. (2004) provided a unique picture of the population dynamics of the species over a period of 60,000 years. Their analysis of 685 bp of mtDNA, control region sequences recovered from 442 archaeological bison bones, combined with AMS dating of 220 of those samples, indicate that a very large and genetically diverse population of bison lived across Beringia until approximately 37,000 years ago. At this point a rapid decline in genetic diversity occurred. The timing of this reduction in diversity does not coincide with either the climate events of the LGM or the Pleistocene-Holocene boundary. It also predates the evidence of human arrival in Beringia by some 15,000 years. The authors point out that this date does correlate with the onset of the last interglacial cycle (MIS 3 which dates to ~60–25 KYA), and that it may have been the combination of the loss of the steppe-tundra during the warm period, followed by the cold and dry conditions of the next glacial period that stressed the bison populations. They also note that the date for population decline in bison coincides with local extinction events in Alaskan brown bears and hemionid horses (wild asses) that may have been similarly affected by climate changes (Shapiro et al. 2004). So, clearly, some large mammal species were already stressed and in decline, if not already extinct, prior to the arrival of humans in the New World.

Despite the evidence for extinction and/or significant population decline prior to human arrival in the Americas, the archaeological faunal record suggests that several other species, such as the woolly mammoth and the horse, did become extinct in North America between 15,000 and 13,000 BP (Grayson 2007). In addition to the arguments for rapid overkill or the "blitzkrieg" by human hunters, it has recently been suggested that a major extraterrestrial collision over North America occurred at 12,900 BP. Firestone et al. (2007) suggest that this event contributed to the Younger Dryas cooling, and that the thermal pulse caused by the event would have resulted in major biomass burning further driving North American megafaunal extinctions. Interestingly though, aDNA evidence challenges the timing of these associations. In their analysis of ancient sediment DNA (sedaDNA) recovered from permafrost cores taken from the Yukon flats of interior Alaska, Haile et al. (2009) found that both woolly mammoth and horse DNA could be recovered from sediments that were securely dated to 10,500 years BP. This finding means that humans and megafauna overlapped for several thousands of years in Alaska, and that at least two species survived the environmental impacts that would have been brought on by the proposed extraterrestrial impact (Haile et al. 2009).

While these aDNA studies provide strong evidence that megafaunal extinctions may not be so easily associated with either human arrival or climate change, another recent aDNA study shows how we might be underestimating the impact of human arrival on native species. Ancient DNA analyses of penguin remains have identified the presumably human mediated extinction of a previously unrecognized penguin species in New Zealand (Boessenkool et al. 2009). It had been assumed that the endangered New Zealand yellow-eyed penguin (*Megadyptes antipodes*), also known as the *hoiho*, is a declining remnant of a previously abundant, native species. This penguin is a New Zealand icon, and is considered a *taonga*, or treasure, by New Zealand Maori. Therefore, they are of significant conservation concern in New Zealand. In an attempt to assess changes in genetic diversity in the New Zealand *hoiho* population over time, Boessenkool et al. (2009) studied the remains of prehistoric, historic, and modern DNA samples. They undertook an analysis of 402 bp of mtDNA, control region sequences and found a previously unidentified genetic split in *Megadyptes*. Interestingly, all but three of the aDNA samples collected from archaeological sites on the South Island of New Zealand that dated prior to AD 1500 formed a clade distinct from all historic and modern samples. They also found that none of these ancient haplotypes were found in any modern *hoiho* populations. This genetic evidence, combined with a further morphological study of the ancient remains, indicated that the specimens represented a previously unrecognized species of *Megadyptes* that became extinct sometime around AD 1500. The apparent extirpation of this native species, which the researchers named *Megadyptes waitaha*, allowed for the range expansion of *Megadyptes antipodes* from the subantarctic islands into the South Island of New Zealand.

Human Impacts on Plants and Animals—Domestication

The Neolithic Revolution was a major event in human history. While it was once believed that the Fertile Crescent was the single center of plant and animal domestication some 10,000 years ago, it is now clear from research in a range of fields that domestication independently occurred in numerous locations around the world and at various points in time. Major regions of early domestication include southwestern Asia/the Near East, Mesoamerica, South America, East Asia, and New Guinea (Zeder et al. 2006).

Archaeologists have long studied the emergence and spread of domesticated plants and animals. By investigating morphological variation in domesticated species and change in the human relationships with those species, many issues can be addressed. This includes, for example, penning of animals, changes in the demographic patterns of faunal remains, the construction of irrigation channels and field boundaries, the clearing forests, and the recovery of artifacts used in these activities, such as hoes and sickles (Colledge 1998; Delcourt et al. 1986; Denham et al. 2003; Golson 1977; Piperno 1993; Piperno et al. 1991).

Geneticists have also been contributing to studies of plant and animal domestication since the earliest development of molecular anthropology. Some of the first immunological studies (Sarich 1977; Seal et al. 1970) clarified issues regarding which species were domesticated. As will be discussed in further detail below, these studies showed that dogs were clearly derived from grey wolves rather than other canid species, such as jackals or coyotes. Since then, the development of new DNA technologies has allowed researchers to study most of the domesticated animals and some of the important plant species, to understand their economic development rather than specifically to address the domestication process itself. To date, there have been aDNA studies addressing the domestication of dogs (Boyko et al. 2009; Cruz et al. 2008; Germonpré et al. 2009; Leonard et al. 2002; Savolainen et al. 2002; Savolainen et al. 2004; Vilà et al. 1997; Vilà et al. 1999; vonHoldt et al. 2010), cattle (Achilli et al. 2009; Beja-Pereira et al. 2006; Bollongino et al. 2008; Bradley et al. 1996; Edwards et al. 2007; Geigl 2008; Götherström et al. 2005; MacHugh et al. 1997), sheep (Hiendleder et al. 2002; Meadows et al. 2007; Pedrosa et al. 2005; Tapio et al. 2006), goats (Luikart et al. 2001; Pereira et al. 2009), horses (Achilli et al. 2012; Cai et al. 2009; Jansen et al. 2002; Ludwig et al. 2009), donkeys (Beja-Pereira et al. 2004; Kimura et al. 2010), pigs (Giuffra et al. 2000; Larson et al. 2005;

Larson and Albarella et al. 2007; Larson and Cucchi et al. 2007), chickens (Liu et al. 2006), cats (Driscoll et al. 2007), turkeys (Speller et al. 2010), water buffalo (Kumar et al. 2007; Yindee et al. 2010), reindeer (Røed et al. 2008), and camelids (Kadwell et al. 2001).

Similarly, the genetics of plant domestication have been addressed for a wide variety of species. Those species most likely to interest archaeologists include maize (Doebley 2004; Jaenicke-Després et al. 2003; Zizumbo-Villarreal and Colunga-GarcíaMarín 2010), rice (Izawa et al. 2009; Khush 1997; Kovach et al. 2007; Sang and Ge 2007; Sweeney and McCouch 2007; Tang and Shi 2007; Zhang et al. 2009), gourds (Clarke et al. 2006; Erickson et al. 2005), olives (Breton et al. 2009; Elbaum et al. 2006), grapes (Arroyo-García et al. 2006; Lopes et al. 2009), taro and yams (Matthews 1990; Terauchi et al. 1992; Yen 1993; Yen and Wheeler 1968), and bananas (Carreel et al. 2002). Research has even looked at genetic evidence for the domestication of yeast (Liti et al. 2009), providing important information about the production of key food items such as beer, bread, and wine.

Plants

Plant remains suitable for aDNA analysis are seldom recovered from archaeological sites, so this is an area where one might think that genetic analyses would be particularly useful. While this is true to a certain extent, the lack of ancient samples suitable for aDNA analyses does have significant drawbacks. As with all studies of modern genetic variation, there are significant difficulties when trying to interpret the variation seen today as it relates to issues about past events. This is even more of a problem for plants than it is for animals. Plant genetics are extremely complex, and while studies of uniparentally inherited markers are available for them, these are not as straight forward to analyze as they are in animals. Rather than being diploid, or having two copies of each chromosome with each one being inherited from one or the other parent, plants are commonly polyploid, meaning they have multiple copies of each chromosome. In addition to the complex nature of inheritance in plants, it is much more difficult to keep domesticated plants reproductively isolated from wild strains, particularly early in the process of domestication. Consequently, introgression (or back breeding with wild varieties) is much more common. Furthermore, the genomes of plants tend to be highly variable and much larger than mammalian genomes. As a result, the most common method of analysis of genetic variation in plants has been through the use of a technique called Amplified Fragment Length Polymorphisms, or AFLP (Meudt and Clarke 2007). An AFLP analysis is similar to an RFLP (see chapter two), but it involves a PCR step that amplifies the cut up fragments of DNA. The main point is that like RFLP analyses, this method provides data reflecting patterns of variation that are less specific than DNA sequence data. This is necessary to get a general picture of overall genetic variation when one lacks prior information regarding specific variation.

Initial studies of AFLP variation in a range of domesticated plant species, including einkorn wheat, emmer wheat, and barley apparently indicated that each of these species was derived from a single and rapid domestication event. This genetic evidence contradicted archaeobotanical data, which suggested that plant domestication was a much slower, complex process that happened in a number of locations around the world and not only in the Fertile Crescent region (Brown et al. 2009). It was later discovered (Allaby et al. 2008) that the phylogenetic analysis of AFLP data would inevitably result in a monophyly, or the indication a single common ancestral lineage for all samples, and therefore a presumed single domestication event, purely by the nature of the data itself. For this reason, Allaby and colleagues designed a simulation to explore what various models of plant domestication might look like in terms of AFLP or RFLP data. They started with a number of "wild" populations consisting of 10,000 plants, within which there was a range of variation. They created a number of "domesticated" populations and randomly chose a subset of the wild samples, which they let grow to various sizes. They then allowed only some of the plants to cross with other domesticated and wild populations. They sampled the variation every twenty-five generations and surprisingly found that regardless of the number of domestication events they created, "the proportion of monophyletic trees increases over time, and all trees appear monophyletic after a number of generations approximately equal to twice the population size" (Allaby et al. 2010:154–155). In other words, even when they created multiple known domestication events and then sampled from them, all phylogenetic reconstructions of the total variation appeared to indicate only a single domestication event. Therefore, while analyses of AFLP data are useful for identifying and studying genetic variation, they are not particularly useful for reconstructing the numbers of domestication events that have involved plant species. The use of other methods, however, such as microsatellite markers (see Box 2.2) and the actual sequencing of particular genes associated with traits related to domestication, have provided data that allow researchers to reliably identify multiple domestication events. Several studies now suggest a scenario for multiregional domestication of a range of species, as well as multiregional selection for particular traits. Therefore, the genetic and archaeobotanical studies are now producing a unified view of the complex process of plant domestication (Brown et al. 2009).

In addition to changes in seed size due to human selection for larger seeds, one of the key indicators of plant domestication—grasses in particular—is the evolution of a trait known as nonshattering. Seed shattering is advantageous, and in fact necessary, for wild seed dispersal. When the seed-containing spikelet reaches maturity, it becomes brittle and easily breaks off from the ear. If humans are going to harvest the seed, they need it to stay attached to the shaft until they collect it in order to collect reasonable amounts. Thus, the process of selection for nonshattering plants may have been an unintentional but inevitable selective pressure during the process of domestication.

Researchers have now identified that a number of SNPs (see Box 2.2 for a general discussion of SNPs) may influence shattering in rice, and they have further identified many other mutations that were clearly selected for during the early process of rice domestication. These include grain size and shape, color, stickiness, and fragrance (Kovach et al. 2007). Similar studies on other domesticated plants are providing new genetic data and tools for addressing issues that are of interest to archaeologists. Of course, the analysis of well-preserved plant remains recovered from the archaeological record increases researchers' ability to address more questions about the past. Recent studies include aDNA analyses of olive pits (Elbaum et al. 2006), gourd rind (Erickson et al. 2005), and maize cobs (Jaenicke-Després et al. 2003).

Animal Domestication

Given the high frequency with which faunal remains are recovered from archaeological sites, they have been the focus of both archaeological and aDNA approaches to understanding animal domestication and husbandry. Genetic studies of living domesticates have also been widely used to address these issues. As has long been recognized, domestication is a process rather than an event. It is, therefore, important in the synthesis of archaeological and molecular data to realize that neither archaeological nor molecular data can address all aspects of the process. The archaeological record can reveal the early manifestations of animal husbandry in the form of penning or accumulations of dung. It can also address this problem by examining the osteological consequences of confinement and selective breeding, such as a reduction in size or alterations in horn morphology. Genetic data, by contrast, are poorly suited to access these phenomena. Studies of modern DNA and those of aDNA are best suited to address different aspects of domestication and animal husbandry, so we discuss them separately.

Modern genetic data are employed to describe the relationships between populations of domestic animals across geography, the timing of divergence of domesticates from their wild ancestors, and of different domesticated populations from each other. The use of molecular data to reconstruct the past in this way is complicated by several factors. First, the patterns of genetic diversity created prehistorically are now, in many species, obscured by historic and modern movement of animals across the landscape. The process is now almost worldwide in many cases due to our ability to airlift frozen sperm and embryos. Additionally, the relatively recent and severe artificial selection in the production of task specific breeds, such as show dogs or fighting chickens, has exacerbated the erasure of signals from prehistory. Second, as discussed in chapter two, assigning absolute dates to the separation of populations is difficult and usually necessitates the inclusion of a wide range of dates. In addition, even if the date of the genetic separation of a domestic species from its wild ancestor is reconstructed with some precision, stochastic processes of DNA sequence evolution result in a date that is older than any archaeologically visible indications of husbandry, or even any changes in human behavior that might have initiated the domestication process. The date on the node separating the incipient domestic population from the wild population will be older than the time at which the population was first confined. This pattern is made worse by the tendency of domestic animals to interbreed with wild populations and thereby introduce new genetic lineages into the domestic gene pool.

In several cases including cattle, sheep, and pigs, the same broadly distributed wild species have undergone domestication in different places, at different times, and from different regional populations. Naturally, subsets of those regional populations entered into domestication processes long after they shared any common ancestor. Despite these independent domestication processes, however, the descendant domesticates are (almost always) able to interbreed freely, and this can result in the wide dispersal of genes originally found in a restricted geographic area.

Dog Domestication

The domestication of dogs is an interesting but contested subject. While the species ancestral to domesticated dogs has been long debated, it is now clear and almost universally accepted that dogs were domesticated from gray wolves and that other canid species have not significantly contributed to the modern domesticated dog, gene pool (Tsuda et al. 1997). Vilà et al. (1997) published a study of mitochondrial, control region diversity of a large sample of domesticated dogs and gray wolves. They found a high level of genetic diversity within domesticated dogs that was similar to that among gray wolves. By calibrating a molecular clock (see chapter two) with a paleontologically dated divergence between wolves and coyotes at one million years ago, Vilà et al. (1997) concluded that domesticated dogs may have diverged from gray wolves approximately 135,000 years ago. This date is in stark contrast with the archaeological record, which provides no evidence of dog domestication until some 120,000 years later (Sablin and Khlopachev 2002). Vilà et al. (1997) reached their conclusion by assuming a single origin of the domesticated dogs in the major clade. In contrast, Savolainen et al. (2002) assume that dogs may have been semi-independently domesticated from local wolf populations across Asia, and they analyze mtDNA to calculate the beginning of the process of domestication. Their results suggest this date to have been about 15,000 years ago, which is much more consistent with the archaeological record. Vilà and his colleagues (Wayne et al. 2006) continue to argue for an ancient domestication of dogs, although not quite as ancient as the original estimate of 135,000 years. They contend that despite their genetic isolation from wolves, dogs need not have exhibited any morphological changes during their long association with humans and only did so after sedentary life ways emerged and changed the nature of the selective pressure on dogs.

Nuclear DNA data provide a somewhat different picture than do the mtDNA data. von Holdt et al. (2010) examined 48,000 SNPs across the nuclear genomes of domesticated dogs and grey wolves. They found that dogs share more of their nuclear variation with Middle Eastern grey wolves than they do with the wolves of eastern Asia. The contrast between the nuclear and mitochondrial results is consistent with both the likelihood that dogs were domesticated in multiple locations and with significant interbreeding between domestic and wild populations.

Furthermore, it is likely that modern, domestic dog breeds do not entirely represent the genetic diversity of more ancient lineages. Studies of mtDNA variation in European, Neolithic dogs have found high frequencies of haplogroup C in dogs from France (Deguilloux et al. 2009), Sweden (Malmström et al. 2008), and Italy (Verginelli et al. 2005). Haplogroup C is found in fewer than 5% of modern European dogs, suggesting that there has been significant reworking of the European dog population since the Neolithic. Likely, this is a phenomenon not constrained to Europe.

Cattle Domestication

Modern domesticated cattle comprise two distinct, though completely interfertile groups—the flat-backed, taurine cattle (*Bos taurus*) and the cericothoracic-humped, indicine cattle (*Bos indicus*). It is generally accepted that the wild ancestor of cattle was the auroch, *Bos primigenius*, which is now extinct but was widely distributed throughout Asia, northern Africa, and Europe (Zeuner 1963). It is also generally accepted that taurine cattle were domesticated from the Eurasian subspecies of auroch, *Bos primigenius*

primigenius. Evidence suggests a reduction of size in cattle from the middle Euphrates river basin 8500 BP (Peters et al. 1999) followed by a change in the sex ratio with females becoming a larger proportion of the population and an absence of older individuals. Indicine cattle were domesticated from the South Asian subspecies *Bos primigenius namadicus* (Clutton-Brock 1989), possibly in Baluchistan during the 5th millennium BC (Bökönyi 1997; Meadow 1993). Still contentious is the role that the northern African subspecies of auroch, *Bos primigenius opisthonomus*, played in the development of modern cattle breeds. Perhaps it was the ancestor of a third independent domestication in North Africa (Bradley et al. 1996; Gautier 1984; Grigson 1991; Wendorf and Schild 1998). Or, it may have contributed wild genes into the domestic taurine populations moving into Africa some 7,000 years ago (Smith 2005). Perhaps it made no contribution at all (Clutton-Brock 1989; Wetterstrom 2001).

Mitochondrial DNA data are concordant with separate domestication processes and separate ancestral subspecies that resulted in taurine and indicine cattle. Analyzing six European and three Indian breeds, Loftus et al. (1994) found that not only did the European and Indian cattle segregate completely into two separate clades, but that the MRCA of the two clades lived anywhere from 200,000 to one million years ago. Likewise, Y chromosome variation indicates a distinction between taurine and indicine cattle (Bradley et al. 1994; Hanotte et al. 2000).

Interestingly, analyses of mtDNA preserved in archaeological, European auroch indicate no direct maternal relationship between the ancient European auroch and modern European cattle (Edwards et al. 2007). Instead, modern European cattle appear to be descended from the Near Eastern auroch. The Y chromosome data, however, indicate a close genetic relationship between ancient European auroch and modern European cattle (Götherström et al. 2005). Combining these data suggests that domestic cattle expanded into Europe from the Near East, and that while they did so, female cattle frequently interbred with wild male auroch such that the descendant population retained its Near Eastern maternal ancestry but had significant paternal input from the European auroch population. These data provide no indication about the extent to which the wild-domestic mating was managed or encouraged.

Sheep Domestication

Domestic sheep belong to the genus *Ovis*, which also contains at least five and perhaps as many as seven wild species of sheep (Geist 1991; Lydekker 1912; Mason 1996; Shackleton 1997). Wild caprines are found in the archaeological record prior to the Neolithic but represent only a very small part of the human diet (Henry et al. 1985; Legge 1996; Uerpmann 1987). During the Aceramic Neolithic, faunal assemblages at many sites in southwestern Asia transition from being comprised largely of wild, noncaprine mammals to ones in which caprines represent the majority of the fauna (Legge 1996). However, among these sites, as well as those that demonstrate a low frequency of caprines, the presence of the species cannot be taken as a direct indication of domestication because the remains could be those of wild animals that were hunted or confined but not domesticated (Bökönyi 1977; Clutton-Brock 1971; Clutton-Brock and Uerpmann 1974; Legge 1975, 1977). During the Aceramic Neolithic period, on the other hand, caprine body size decreased significantly, which is particularly striking given that there is an observed stability in caprine body size during the Paleolithic, Mesolithic, and proto-Neolithic (Helmer 1985; Meadow 1984, 1993).

These kinds of archaeological data do not allow a reliable estimation of the identity of the specific species ancestral to modern domesticated sheep. This is a task for which genetic data are better suited. Mitochondrial DNA data, as well as direct examinations of the numbers and sizes of chromosomes (the karyotype), indicate that the Asiatic mouflon (*Ovis oritentalis*) is the likely ancestor of all modern domestic sheep (Bruford and Townsend 2006; Bunch et al. 1990; Hiendleder et al. 1998; Hiendleder et al. 1999; Lyapunova et al. 1997; Meadows et al. 2011). Variation in the mitochondrial genome falls into five haplogroups, A through E. The MRCA of the two most common haplogroups, A and B, dates to about 920,000 years ago. The two most closely related haplogroups, C and E, date to about 260,000 years ago (Meadows et al. 2011). These ancient divergences between haplogroups suggest that sheep were domesticated several times; although, it is possible that these haplogroups were introduced into domestic sheep populations through introgression from wild sheep. The results suggest that the development of modern domestic sheep was a complex, geographically diverse, and lengthy process.

Pig Domestication

Like cattle and sheep, the genetic data from pigs indicate that they were also domesticated in multiple locations. Wild and domestic pigs are both members of the species *Sus scrofa*, which originated in western ISEA (Larson et al. 2005) and later spread across India into eastern Asia, Eurasia, and western Europe. Previous evidence suggested that European pigs were descended from Near Eastern pigs spread by the expansion of Neolithic farmers from the Near East. Mitochondrial DNA evidence, though, suggests instead that pigs were domesticated from at least two lineages of wild European boar (Larson et al. 2005). These European domestications are in addition to those that occurred in eastern Asia (Giuffra et al. 2000; Kijas and Andersson 2001), the Near East, India, and Sahul (Larson et al. 2005). Interestingly, examinations of mitochondrial variation of wild boar populations indicate that there are few wild boar across Eurasia that are descended from feral domestic pigs (Larson et al. 2005).

Ancient Genetic Variation in Goats—DNA from Goat Skin Clothing

Most molecular anthropology directed at understanding domestication employs analyses of DNA extracted from tissue samples obtained from living animals or the bones and teeth of archaeological fauna. Sometimes, however, the artifacts constructed from secondary animal products can be used. A leather legging, discovered in the wake of a retreating glacier in the Swiss Alps, is an example of such an artifact (Schlumbaum et al. 2010). Morphological analyses identified that the legging was likely to be made from the skin of a goat, although it differs somewhat from modern goatskin. Extraction, amplification, and sequencing of mtDNA from the specimen revealed it to belong to goat mtDNA haplogroup B, a now relatively uncommon clade (Schlumbaum et al. 2010). The vast majority (91%) of modern goats belong to haplogroup A, which has a worldwide distribution (Naderi et al. 2008). Haplogroup B, the next most common haplogroup, has been identified in about 6% of modern goats and is primarily found in Asia and sub-Saharan Africa. However, a single individual has also been found in Greece. Haplogroups C, D, E, and F are found only at very low frequencies in Europe. Previous work has suggested that haplogroup B was in modern European goats as a consequence of a relatively recent expansion of goats originating in the Near East and beginning only about 2,000 years ago (Luikart et al. 2001). However, there is no archaeological evidence of trade from the Near East, which would suggest that the leather might have arrived in the Swiss Alps as an already-manufactured piece, and there are bone artifacts with polish recovered from nearby and contemporaneous sites implying a local leather working industry (Schlumbaum et al. 2010). Thus, the authors suggest the artifact was made from an animal that was likely obtained locally. When considered with the results from the mitochondrial sequencing of nineteen early Neolithic goats from southern France (Fernández et al. 2006), which reveal eight members of haplogroup A and eleven members of haplogroup C, it does suggest that the goats in Neolithic Europe were a much more genetically diverse population than they are today.

Horse Domestication and Human Selection for Coat Color

In addition to reconstructing population relationships, genetic data can be employed to understand the changes in domesticated animals as a consequence of artificial selection. In a study of eight mutations across six genes responsible for variation in coat color in archaeological horse remains from across Siberia and eastern and central Europe, Ludwig et al. (2009) found dramatic changes over time. They found no variation among the Pleistocene horses suggesting that all of the horses were bay or dun-bay. By the early Holocene, however, the horses from the Iberian Peninsula were divided between black and bay colored as were eastern European horses dating to between 7,000 and 6,000 years old. It has been argued that horses were probably not domesticated until about 5,500 years ago (Outram et al. 2009), which may mean that these horses were wild. Thus, variation in their coat color could have been the result of selection caused by increased forestation or by the repopulation of the region from elsewhere following the retreat of the glaciers. After 5000 BP, there was increasing variation in coat color in the horses across most of Europe; although interestingly, it was not until the Middle Ages that any coat color change was seen among Spanish horses. With improvements in aDNA recovery and sequencing technology, we are hopeful that more studies of this nature will reveal the patterns of both artificial and natural selection on plants and animals as they underwent domestication and subsequent adaptation.

DNA and the Identification of Faunal Remains—Subsistence

The remains of closely related species are frequently difficult to distinguish morphologically, particularly when fragmentary (Gobalet 2001). However, the accurate identification of the species in a faunal assemblage is often critical to interpretation. When preservation is adequate, a molecular approach can significantly enhance the chances of precise species identification.

Throughout the Pacific Ocean, fish are a major component of faunal assemblages, reflecting their importance in prehistoric Pacific diets. Typically, fish are difficult to distinguish below the family level, but fish within a family can exhibit widely divergent habitats and thereby require substantially different fishing techniques. In Aitutaki, in the southern Cook Islands, fish of the family Serranidae comprise as much as 40% of some of the island's archaeological fish assemblages. Members of this family, however, occupy habitats from inshore lagoons to outer reef and deepwater ecosystems. The difficulty in identifying these fish to the species level has prevented a determination of their origins and, thereby, the necessary fishing technology and practices necessary to procure them. It also prevents the detection of any changes in the fish diet over time. By sequencing a portion of the mitochondrial genome from fish bones, Nicholls et al. (2003) were able to determine that the majority of the

Serranids recovered from the site were inshore species and, therefore, obtaining them did not require outer-reef or deepwater fishing.

In another application of molecular species identification to the understanding of prehistoric human subsistence, Barnes et al. (2000) employed aDNA analysis to determine whether the geese found at two sites in Lincolnshire, United Kingdom, were wild or domestic. Among six possible species that cannot be distinguished morphologically, they found only domestic geese at Vicars Court (seventeenth–nineteenth centuries) but both domestic geese and wild, pink-footed geese at Flixborough (seventh–twelfth centuries), indicating that the residents engaged in both animal husbandry and wild fowling.

DNA Analysis of Animal Products as Evidence of Trade and Exchange

The geographic origins of biological products involved in trade and exchange networks can sometimes be identified using molecular techniques. Akin to the identification of lithics manufactured from nonlocal sources, simply identifying the nonlocal species is a useful step in tracing trade networks. Furthermore, phylogenetic analyses of DNA extracted from traded animal parts can sometimes illuminate their geographical and, thereby, cultural origin.

Catfish Remains in Turkey

Arndt et al. (2003) studied catfish (*Clarias*) remains from the Roman and early Byzantine components of the site Sagalassos in Turkey. Catfish live in the Levant and in Egypt, but they do not occur naturally near Sagalassos. By analyzing mtDNA sequences from the Sagalassos catfish assemblage and comparing them with those of modern catfish, the researchers were able to identify the lower Nile as their source. Given the distances separating the locations, it is assumed that the fish must have been processed either by drying or by smoking before they were traded. This link is consistent with other archaeological evidence of trade and exchange between the people of Sagalassos and communities located in the Nile valley. The result, however, highlights the potential for further aDNA analyses of fishbone in general, and *Clarias* remains in particular, for tracing the widespread, ancient trade of Nilotic fish throughout the Mediterranean region.

Importation of Raw Materials or Finished Artifacts?

In a similar study of trade and exchange in the American Southwest, Borsen et al. (1998) used mtDNA analysis to investigate an Anasazi artifact recovered in 1954 from a cave in the state of Utah. The artifact, which was unique in form and construction, consisted of macaw feathers, a tassel-eared squirrel (*Sciurus aberti*), and yucca, and it dated to 920 +/– 35 BP. Borsen and colleagues wanted to determine the origins of the various components in order to better understand the trade process. Specifically, researchers wanted to know if the completed article was traded from outside the Southwest or if the various components were traded and then assembled in the region. Scarlet macaws are tropical birds, endemic to Mexico, and their range does not extend as far north as the American Southwest. Therefore, at the very least, the macaw feathers were part of the known, large-scale, trade network that moved living birds and feathers northward. In contrast, the range of the tassel-eared squirrel extends throughout the pine forests of Mexico and the Southwest. To determine the origin of the squirrel and gain insight into where the artifact was manufactured, Borson et al. (1998) analyzed part of the mtDNA sequence from the squirrel pelt. They compared the results with sequences from tassel-eared squirrels across their modern range. The sequences showed that, while the macaw feathers must have been trade goods, the squirrel pelt was of local, Southwestern origin. It remains conceivable that the feathers were bound with yucca fibers in Mexico before being traded north, but the Borsen and colleagues contend that it is more likely that all the manufacturing was undertaken in the Southwest, combining imported feathers with a local pelt and local yucca.

Feather Cloak Construction in New Zealand

In the early 1800s, New Zealand Maori began constructing flax fiber cloaks completely covered in the feathers of both native and nonnative bird species. These cloaks had significant cultural status, and those of the highest prestige were constructed from kiwi (*Apteryx* spp.) feathers. To investigate which of the five species of kiwi were used in cloak manufacturing, what the geographical origins of the kiwi were, and which sex was used, Hartnup et al. (2011) recovered mtDNA sequences from 849 feathers from 109 cloaks. They assayed them for sex chromosomes to determine the sex of the bird for ninety-two feather samples. They found that the vast majority of the feathers were from the North Island brown kiwi, and that about 15% of the cloaks were constructed from feathers of different geographical origins, suggesting either extensive trade in kiwi feathers or kiwi hunting across a wide geographic range. Furthermore, over half of the cloaks were made of feathers derived only from kiwi populations in the eastern North Island, suggesting that this may have been a region of particularly active cloak manufacturing. Finally, the results showed a slight, but not statistically significant, excess of male feathers, which if reflective of the real pattern may been the result of the increased vulnerability of males while incubating eggs.

The ability to recover DNA from artifacts is one area where the collaboration between archaeologists and molecular anthropologists can be particularly fruitful. Where much of the research on domestication and extinction events has been driven by biologists interested in animal husbandry or broader evolutionary questions regarding extinction events, we suggest that the analysis of artifacts is an area where the research is best directed by archaeologists and museum staff. Unfortunately most aDNA analysis is destructive. Yet, there are several groups of researchers working on methods that are nondestructive or minimally invasive (Rohland et al. 2004; Thomsen et al. 2009). We hope that these approaches, combined with the advantages of NGS for aDNA studies will allow for further research into the analysis of ancient artifacts and other precious archaeological remains.

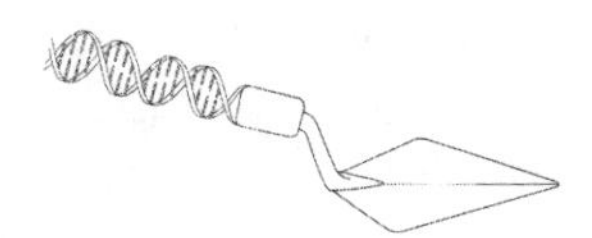

CHAPTER EIGHT

Individualization and Other Applications of Ancient DNA

In the previous chapters, we discussed the application of aDNA analyses to address questions about hominin evolution, modern human population expansions, and the processes of plant and animal domestication. These are questions that relate to large-scale, anthropological issues. Many archaeologists, however, may want information relating to a specific historic event, a specific archaeological site, or even a specific individual. Ancient DNA analyses can also provide valuable information to address some of these more proximate questions and allow for the reconstruction of information about individuals that cannot be determined, or determined with any certainty, using traditional archaeological or osteological methods. In this chapter we discuss some of the aDNA methods used to address issues such as health and disease or the identification of sex and other personal characteristics of individuals. We also present some interesting case studies that demonstrate the application and value of these methods.

SEX DETERMINATION

Typically, it is difficult or impossible to determine the sex of fragmentary or subadult skeletal human remains. In circumstances when the sex of particular individuals affects the interpretation of an archaeological site, attempts to determine sex with genetic methods are often warranted. Molecular approaches to sex determination typically follow one of two related strategies; although because both necessitate the recovery of nuclear DNA, they are likely to be successful on only a subset of those

DNA for Archaeologists by Elizabeth Matisoo-Smith and K. Ann Horsburgh, 155–169.

samples from which mtDNA can be recovered. One approach attempts to recover a small fragment of the Y chromosome that does not have an equivalent on the X chromosome (Cunha et al. 2000; Stone et al. 1996). If the fragment can be recovered, then a male is indicated. However, the failure to recover the fragment can indicate either that the individual was female, or that the DNA is too degraded for analysis, a result known as allelic dropout. The other strategy targets a section of the amelogenin gene, which is present on both the X and the Y chromosomes (Mannucci et al. 1994). The section of the gene is slightly longer on the Y chromosome than it is on the X chromosome (Daskalaki et al. 2011) allowing the two fragments to be distinguished. As every individual has at least one X chromosome, the failure to recover either fragment indicates a lack of DNA preservation.

While much work to date has been directed at verifying the reliability of molecular sex determination (Arnay-de-la-Rosa et al. 2007; Faerman et al. 1995; Schmidt et al. 2003; Stone et al. 1996; Vernesi et al. 1999), broader anthropological questions are also being directly addressed this way. Excavations at a temple at the Aztec site Tlateloco, Mexico, uncovered the remains of thirty-seven subadults and six adults who seem to have been sacrificed to the Aztec god of wind and rain. The subadults were largely intact, but the adults were significantly fragmented. To better understand the ceremonies in which these people were sacrificed, a team attempted to determine the sex of the recovered remains (De La Cruz et al. 2008). The site is approximately 550 years old, and the DNA appears to have been remarkably well-preserved. This allowed the researchers to recover nDNA and amplify X and Y chromosome fragments of 192 bp and 158 bp respectively. Interestingly, the vast majority of the individuals appear to have been male, which is consistent with a previously developed idea that sacrificial victims were chosen as living embodiments of the male god. Despite the interesting aspects of this study, it is worth noting that aDNA fragments would typically be sequenced, or cloned and sequenced, but that in this study, they were not. Given the widespread availability and low cost of sequencing, it is not clear why it was not undertaken. This step is particularly important and would have provided an opportunity to investigate possible contaminating sequences that would not have been visible to presence/absence or RFLP (see chapter two Box 2.2) scoring. Nonetheless, the researchers did undertake independent replication of results in another lab, and thus, there can perhaps be some degree of confidence in the data and their anthropological implications. We would, however, still suggest caution because the researchers did not follow the most stringent aDNA protocols and have not proven, beyond doubt, that they did not amplify modern DNA contamination.

Frequently, the value of secure sex determination of human remains in an archaeological site depends on having results from a significant proportion of the population (Cappellini et al. 2004; Cunha et al. 2000; De La Cruz et al. 2008; Faerman et al. 1997), which has often been unattainable because of poor DNA preservation. We are optimistic that, with the development of new recovery techniques associated with NGS technology, nDNA recovery from more sites, and from more individuals within a site, will be possible.

Infectious Disease

In addition to the more typical archaeological samples from which DNA is recovered, such as animal bones, and occasionally plant remains, many attempts have also been made to analyze the DNA of infectious agents and, thereby, characterize the illnesses of long-dead people. Traditional paleopathological research can be frustrated by the presence of skeletal lesions that are consistent with, but not diagnostic of, infection with specific pathogens. Or, researchers may suspect disease based on known historic events but be unable to recognize corresponding changes in the skeleton. In these circumstances, the analysis of DNA from pathogens that have invaded the bone tissue is attractive. This approach has been attempted with Chagas' disease (Aufderheide et al. 2004; Fernandes et al. 2008), *E. coli* (Fricker et al. 1997), Hansen's disease, or leprosy (Donoghue et al. 2005; Haas, Zink, Molńar et al. 2000; Matheson et al. 2009; Taylor et al. 2000; Taylor et al. 2006), influenza (Stevens et al. 2004; Taubenberger et al. 1997), malaria (Sallares and Gomzi 2001; Taylor et al. 1997), the plague (Drancourt et al. 1998; Drancourt et al. 2007; Gilbert et al. 2004; Kacki et al. 2011; Raoult et al. 2000; Schuenemann et al. 2011; Wiechmann and Grupe 2005), syphilis (Bouwman and Brown 2005; Kolman et al. 1999; von Hunnius et al. 2007), and tuberculosis (Baron et al. 1996; Bathurst and Barta 2004; Donoghue et al. 1998; Fletcher, Donoghue, and Holton et al. 2003; Matheson et al. 2009; Taylor et al. 1996; Zink et al. 2003).

While the controls against contamination have become evermore stringent among mainstream aDNA research, analyses of archaeological pathogens have frequently been conducted under the assumption that they are less susceptible to the introduction of exogenous DNA and that structural qualities of some pathogens allow the preferential preservation of pathogenic DNA. Consequently, much of the work published on the aDNA of infectious diseases has been met with skepticism. A 2008 review of sixty-five papers published between 1993 and 2006, each of which reported the recovery of ancient pathogenic

DNA, found serious flaws in most of the studies (Roberts and Ingham 2008). The paper addressed many issues in aDNA recovery, but the most noteworthy was that almost half of the studies made no reference to the use of a dedicated aDNA laboratory that was physically separated from other molecular biology labs. In addition, only 11% of papers reported the recovery of mtDNA from the samples analyzed, which would be expected to be at least as well-preserved as genuinely ancient microbial DNA. Furthermore, many researchers studying ancient pathogenic DNA explicitly rejected cloning as a valuable tool in assessing the validity of their results (see chapter three for a discussion of cloning; Donoghue et al. 2009; Taylor et al. 2009).

In addition to a general failure to adhere to established laboratory protocols for aDNA research, the analysis of ancient pathogenic DNA studies also suffer from difficulties specific to the organisms under investigation. We discuss here the particular example of tuberculosis because it has been the subject of many studies, and its recovery and secure identification remains particularly controversial.

Tuberculosis, caused by *Mycobacterium tuberculosis*, infiltrates the blood and lymphatic systems of infected people. Through the blood and lymph, the bacteria are introduced directly into the bones and surrounding tissues, where they continue to replicate. Therefore, it is possible that significant quantities of pathogen DNA will be present in the bones of the afflicted person and would still remain in the bones postmortem. While this feature of the biology of tuberculosis makes it an ideal target for aDNA research, archaeological bones are not immune from contamination while in the ground. There are more than 100 known species of *Mycobacterium* (Wilbur et al. 2009), and it is likely that there are many more that have not been identified. These species are predominantly found in soil and water and are, therefore, highly likely to come into contact with archaeological remains. Furthermore, while it is true that archaeologists and laboratory workers are unlikely to shed mycobacterial DNA, amplified DNA from previous experiments undertaken in the laboratory is as potent a source of contamination as amplified DNA from other organisms. This fact and the low levels of variability in tuberculosis sequences makes detection of such contamination very challenging.

Despite the difficulties associated with studying the DNA of archaeological pathogens, it remains an alluring area of research. With further study, it will perhaps overcome the problems currently associated with its reliability and credibility. Proteins associated with pathological organisms have begun to receive attention, and it seems likely that a combination of genetic and proteomic research may be a productive approach to molecular paleopathology.

Ancient DNA Approaches to the Identification of Specific Individuals

Analyses of DNA preserved in human remains have been employed to investigate particular questions about historic figures including identifying remains, tracing parentage, and diagnosing disease states. While these approaches are particularly useful for forensic archaeology, law enforcement, or recovery of war dead, the methods can also be valuable for addressing broader anthropological issues.

Individual Identification and Family Relationships—The Romanovs

Historical records of the final days of the Russian royal family, the Romanovs, including Tsar Nicholas II, Tsarina Alexandra, and their five children, are slim. It has been thought that the family, their doctor, Eugeny Botkin, and three of their servants were killed by revolutionaries in Ekaterinburg, Russia, in July, 1918. While transporting the bodies for disposal in a mineshaft, the revolutionaries' truck broke down, and the group was buried in an unmarked grave under a road. Despite significant efforts, however, the narrative was difficult to confirm. When, in 1991, a mass grave was discovered about 30km outside Ekaterinburg, the opportunity was taken to use genetic techniques to attempt a secure identification of the bodies (Gill et al. 1994). Osteological age and sex investigations suggested that there were six adults and three subadult females, suggesting the presence of the Tsar and Tsarina, their doctor, the three servants, and three of the children, leaving one daughter and the only son unaccounted for. Mitochondrial DNA and nuclear STRs (see chapter two for a discussion on STRs) were typed for each of the nine skeletons recovered from the grave, and these results were compared with data from Prince Philip, the Duke of Edinburgh, a maternal relative of the Tsarina, and two maternal relatives of the Tsar. Most interesting were the mtDNA data. One of the adults matched all three subadults and Prince Philip, lending strong support to the conclusion that these were indeed the Tsarina and three of her children. Likewise, the mtDNA of another of the adults was identical to two maternal relatives of the Tsar. The other four adults were shown to be unrelated to each other and unrelated to any of the maternal relatives of the Tsar and Tsarina tested, which is consistent with the presence of the three servants and the physician thought to have been killed with the royal family.

Zhivotovsky (1999) challenged the results largely based on suspicions that the grave was staged and disrupted prior to being properly excavated. Further, Zhivotovsky purported that the adult male thought to be Nicholas might be his brother, who had also been murdered, and that those remains were placed in the grave specifically to provide evidence of Nicholas' lineage. This stratagem would, however, have necessitated furnishing the grave with four relatives of Tsarina Alexandra, as well as those of Nicholas' brother. Zhivotovsky also expressed concern that, while the DNA of the analyzed femora were consistent with the hypothesis that the grave held the Tsar, the Tsarina, three of their children, and the doctor and servants, the rest of the skeletal remains were not. Yet, to suggest that all the analyzed bones are consistent with the Romanov identification, but that the rest of the bones are not, requires either a degree of special pleading or that the team involved in the analysis engaged in a conspiracy to fabricate the impression that the grave contained the remains of the Romanovs.

The results of the original study were further disputed because the quality of the DNA recovered from nine bones was considered "too good" to be genuinely ancient (Knight et al. 2004). Although the level of DNA preservation was indeed excellent, this is not unique among ancient samples (Hofreiter et al. 2004). Nonetheless, based on these suspicions, Knight et al. (2004) analyzed DNA extracted from a finger thought to be from the Grand Duchess Elizabeth Feodorovna, Tsarina Alexandra's sister. They contend that the mitochondrial sequence they recovered differs from that of the remains thought to be Alexandra and her three daughters. They could not, however, account for the differences between the sequence they recovered from the Grand Duchess and that obtained from a blood sample from Prince Philip, a maternal relative of both Elizabeth and Alexandra. It does seem that a lack of identity between the mtDNA from the finger and that of Prince Philip would require significant explanation. Knight and colleagues did, however, recover a sequence that was the same as that recovered from the four bones in question and from Prince Philip, but dismissed it because it occurred less frequently in their DNA extract than another, different sequence (Hofreiter et al. 2004). It is not uncommon in aDNA studies to find that the endogenous DNA remaining in a sample is damaged and occurs at a very low concentration while the contaminating DNA is abundant (Green et al. 2009).

The most recent chapter in the saga of the Romanov family involves reports of the discovery of the two missing children, the Tsarevich Alexei and one of the daughters. In 2007, a second grave was discovered roughly 70 m from the first and found to contain a minimum of two individuals. One was male, one was female, and both were teenagers (Coble et al. 2009; Rogaev et al. 2009). Molecular analyses of these remains, in conjunction with further analyses of remains from the original interment, are all consistent with the original conclusions of

Gill et al. (1994), as well as with the hypothesis that the two individuals in the second location are the two Romanov children missing from the original family grave.

Kinship and Status—The Mycenae Grave Circle B

The analysis of burial sites can provide important information regarding social structure through analysis of grave goods and other archaeological features, such as location and positioning of burials. Changes in burial type and content through time or across a site can, of course, provide key information regarding social status or the development of complexity in social structure through time. Analyses were undertaken of burials from the Bronze Age citadel site of Mycenae, Greece, to address just these types of questions. Bouwman et al. (2008) realized that they could apply aDNA methods to address key questions raised by previous researchers regarding the relationships of individuals within and between two grave circles.

Bouwman and colleagues describe how both grave circles date to the period 1650–1500 BCE and were rich in gold artifacts. They were thus believed to be the graves of elite members of early Mycenaean society. The six burials in Grave Circle A were all shaft burials and represented the remains of nineteen individuals. Grave Circle A had more gold artifacts, including gold facemasks, associated with them than did the burials in Grave Circle B. Not only did Grave Circle B contain simple cist burials, as well as shaft burials, it had many graves that contained a large number of burials with weapons and pottery. A total of thirty-nine individuals were buried in Grave Circle B. The burials appeared to be laid out in four plots within the circle, with each group representing a different period or phase of the site. It was suggested that these might represent family plots. Facial reconstructions of the seven best-preserved burials from Grave Circle B revealed that the facial shape of these individuals fell into three categories—heart-shaped, long, and beak-like faces. One individual, the earliest burial, had a combination of a heart-shaped and a long face.

In order to test the hypothesis that the individuals in Grave Circle A were all members of one ruling, elite family, and that the members of Grave Circle B might represent different family groups, Bouwman et al. (2008) undertook an aDNA study of the skeletal remains. Bone samples from twenty-two individuals were processed to determine if mtDNA and Y chromosome DNA could be obtained. Unfortunately, no nDNA could be amplified, but the researchers did obtain mtDNA from four individuals from Grave Circle B. The mtDNA sequences obtained from two individuals, labeled burials Γ55 and Γ58, and identified as an adult male and female were buried together in a grave. Both had heart-shaped faces, and both had mtDNA that belonged to haplogroup U/K. It was determined, based on these similarities, that these individuals were likely to be brother and sister. Burial Z59, belonged to the Long face group and had a mtDNA haplotype designated as U5a1 or U5a1a, and was therefore not maternally related to the brother and sister burials. The fourth sample for which they obtained an mtDNA sequence was burial A62, and his mtDNA matched the Cambridge Reference Sequence for the region. This meant that he could not belong to either mtDNA haplogroup U/K, U5a1, or U5a1a.

These DNA results indicate that other than the possible brother and sister duo, none of the other burials from which mtDNA was obtained were maternally related. They could, however, be related on the paternal side. The relationships identified by aDNA were consistent with those identified through face shape analysis. While the aDNA results could not reject the possibility that the burials in Burial Circle B were all related, the results obtained for the possible siblings allow for some interesting implications. Burial Γ55 was the only burial in the circle to be buried with a facemask, and burial Γ58, his purported sister, was one of only four females buried in the circle. The fact that she was buried with her likely brother was taken as evidence that she may have held a position of authority due to her birth rather than because she married a man of position. This aDNA result, combined with other archaeological and morphological data, provides some possible insights into the social structure and attainment of status in early Mycenaean society.

Individual Characteristics—Nicolaus Copernicus

Polish astronomer Nicolaus Copernicus was born in Thorun, Poland, in 1473. He spent most of his adult life in Frombork, Poland, where he made most of his astronomical observations and where he served as a priest at the Frombork Cathedral. After he died in 1543, he was buried in the Cathedral where he had served, but his, like most of the tombs in the Frombork Cathedral, went unmarked (Bogdanowicz et al. 2009). Attempts to locate his grave were unsuccessful for years, but renewed efforts beginning in 2004 focused on several skeletons discovered near the altar that had been Copernicus's responsibility during his time at the Cathedral. One skeleton in the group was of particular interest because facial reconstruction suggested that it might have been that of Copernicus. Efforts to locate biological relatives of Copernicus to provide comparative genetic material failed, leaving researchers to look among his belongings in the hopes of recovering biologically useful samples.

They found what they needed in the form of nine hairs caught in the pages of an astronomical reference book owned and used by Copernicus.

Bogdanowicz et al. (2009) extracted DNA from three molars and both femora of the skeleton suspected to belong to Copernicus, as well as from the nine hairs found in the astronomy book. They successfully recovered amplifiable DNA from all five skeletal elements and four of the hairs. Two of the hairs and all the skeletal elements all produced sequences with four unusual mutations in the hypervariable region of the mtDNA, suggesting that they had the same origin, which was likely Copernicus—the owner of the book. At the time of reporting in 2009, the identified mitochondrial sequence, which belongs to an unusual haplotype of the common European haplogroup H, had been reported only three times among Germans and once in a Dane. The likelihood that the skeleton and the hair shared this haplotype by chance is very low. The Y chromosome STRs were recovered from one of the teeth and confirmed that the skeleton was male. Finally, the team was able to type an SNP in the *HERC2* gene, which indicated that the individual most likely had light eye coloration (gray or blue eyes). Interestingly, portraits of Copernicus generally depict him with dark eyes. Bogdanowicz and colleagues, however, point out that this discrepancy may have been due to the printing technique commonly used at the time Copernicus was alive that causes poor replication of light colors. All later images of Copernicus, based on these early prints, may have replicated this dark-eyed image.

Paternity—Thomas Jefferson

In 1802, United States President, Thomas Jefferson was accused of being the father of twelve-year-old Thomas Woodson, the son of his slave Sally Hemings, who had five children. The youngest of Heming's children was Eston Hemings Jefferson who, it was believed, bore a strong resemblance to Thomas Jefferson. Martha Jefferson Randolph, Thomas Jefferson's daughter, contended that some of Sally Heming's children, including Eston, were fathered not by Thomas Jefferson, but by one of his nephews. The nephews, surnamed Carr, were the sons of Thomas Jefferson's sister and would not have the same Y chromosome as Thomas Jefferson. Given the differing genetic implications of these scenarios, they ought to be distinguishable with aDNA.

Unfortunately, there are no living descendants of Thomas Jefferson in the male line because he had no surviving sons. His paternal uncle, Field Jefferson, however, did have living descendants in the male line. To investigate the allegation that President Jefferson was the father of Sally Hemmings's children, Foster et al. (1998) examined the Y chromosomes of five male descendants of Field Jefferson, who should share a

Y chromosome haplotype with Thomas Jefferson, five male descendants of Thomas Woodson, one male descendant of Eston Hemings Jefferson, and three male descendants of the grandfather of the Carr nephews.

Four of the five descendants of Thomas Woodson carry a Y chromosome with a European haplotype, but not that of the descendants of Field Jefferson and, therefore, not one which could have been passed on by President Jefferson. However, the fifth of the Woodson descendants had a Y chromosome with a completely different haplotype, implying that somewhere in the lineage the biological father was not the socially-recognized father of one of the sons. This is commonly known as a non-paternity event, indicating maternal infidelity. The descendants of the grandfather of the Carr nephews carried yet again a different European type Y chromosome that is not seen in either the descendants of Thomas Woodson or of Eston Hemings Jefferson. On the other hand, the descendants of Eston Hemings Jefferson did carry Y chromosomes of the same haplotype as the descendants of Field Jefferson and, therefore, presumably President Jefferson (Foster et al. 1998). As Abbey (1999) points out, President Jefferson's brother Randolph, as well as Randolph's five sons, would all have had the same Y chromosome; hence any one of them could have been Eston's biological father. Foster et al. (1999) readily concede the point, but contend that it is known that Thomas Jefferson and Sally Hemmings were both at Monticello at the time of Eston's conception. They also point out that there is no historical evidence suggesting that Randolph or any of his children were there at the time. Thus, Thomas Jefferson is the most likely father of Eston Heming Jefferson.

Genetic Disease—Abraham Lincoln

It was suspected in the 1960s that United States President Abraham Lincoln suffered from Marfan syndrome, a genetic disorder that might account for his unusual appearance. Recently, however, a new diagnosis has been suggested. Multiple endocrine neoplasia type 2B (MEN 2B) causes many of the same skeletal features as Marfan syndrome, such as the long limbs, large flat feet, small head, and sunken chest. Notably, however, MEN 2B does not cause features of the heart and the eyes that are typical of Marfan syndrome, and which Abraham Lincoln seems to have lacked. Photographs of Lincoln do not show the eyes typical of Marfan syndrome. Further, his well-documented reputation as an outdoorsman suggests that he is not likely to have suffered from the heart defect that prevents Marfan syndrome sufferers from undertaking vigorous exercise. In addition, photographs of Lincoln seem to show the overgrowth of neural tissue in the lips, producing bumps, which is typical of MEN 2B. Finally, an MEN 2B diagnosis is consistent with President

Lincoln's rapid deterioration as documented both in photographs and his own descriptions of his declining health toward the end of his life. It may be that, at the time of his assassination, Lincoln was already dying of thyroid cancer, a classic symptom of MEN 2B.

In order to test this diagnosis, one of the authors of this work, Horsburgh, along with John Sotos, the MD who made the MEN 2B diagnosis, have secured samples of a cloth that was present at Lincoln's assassination that shows evidence of Lincoln's blood, and we hope his DNA. The research is in its early stages, but we are optimistic that we may be able to recover portions of the gene implicated in MEN 2B and, if not, perhaps those markers associated with Marfan Syndrome.

Ötzi, the Tyrolean Iceman

Perhaps one of the most famous cases of individual identification and investigation is that of Ötzi, the naturally mummified remains discovered in September, 1991, in the Tyrolean Oetztaler Alps, on the border between Italy and Austria. Ötzi's body was exposed by melting glacier ice, and initial attempts made by amateurs to remove the remains unfortunately caused major damage to both the body and the site context (Seidler et al. 1992). Forensic experts were then brought in to recover the body, and it was taken to the anatomy department at the University of Innsbruck, where it was stored under refrigeration. Initial analyses indicated that Ötzi was a male, 25–40 years old, who stood 156–160 cm tall, and had a cranial capacity of between 1,500 and 1,560cc. He was wearing clothing that consisted of goatskin leggings, a leather jacket, a cloak of woven grass, and shoes made from bear and deer leather that were stuffed with grass, presumably for insulation. Direct radiocarbon dates indicated that Ötzi died sometime between 5,200 and 5,300 years ago, and initial studies suggested that he died from exhaustion. Though, several years later it was discovered that he had an arrowhead lodged in his back that apparently severed his subclavial artery, most likely causing him to bleed to death (Maderspacher 2008).

The initial genetic analyses of Ötzi were undertaken by Handt and Richards et al. (1994). Eight samples of muscle, connective tissue, and bone were removed from the left hip region, which had been damaged during the original recovery attempt of the body. The initial expectation was that Ötzi's DNA would be relatively well-preserved given that he had been frozen (ideal conditions for DNA and tissue preservation) and had only been exposed or thawed for a brief period. However, initial results from the lab showed that there was very little endogenous DNA remaining in the tissues recovered (Handt and Richards et al. 1994). The team was unable to amplify the amelogenin genes used for sex identification (as described previously in this chapter). They were, however, able to amplify small portions of the hypervariable region of the mtDNA. DNA extraction and PCR amplification of a subsample of bone was undertaken in an independent lab, and identical sequences were obtained, confirming that the results were not due to contamination from within the laboratory. It was observed through cloning a large number of PCR products, that there were several mtDNA sequences being amplified. A number of contaminating sequences were then identified, presumably belonging to those individuals who initially tried to recover the body.

Along with colleagues, Handt and Richards were able to amplify and sequence several small, overlapping fragments of Ötzi's mtDNA genome, generating a total of 354 bp of sequence data. They identified two base changes from the CRS and then compared that sequence to a European database. Ötzi's sequence is not uncommon in Europeans living today and is most common in northern Europeans. It has also been recorded in one Swiss individual and three times from Mediterranean populations. The researchers also collected DNA samples from sixteen individuals living in the Ötztal valley, in which Ötzi was found, and additional samples from several communities located in the broader Alpine region (from both Italian and Austrian sides of the border). The sequence obtained from Ötzi was most closely related to individuals from the Alpine region, but his sequence was not found in the sixteen individuals from the Öetzal valley.

Since the initial research by Handt and Richards et al. (1994), several additional studies have been undertaken to further analyze Ötzi's mtDNA (Rollo et al. 2006), including a recent project that obtained his complete mtDNA genome (Ermini et al. 2008). Researchers have been able to confirm that Ötzi had mtDNA that belonged to the lineage K1* and that it also possessed two mutations that have not been identified in any living European populations (Endicott et al. 2007; Ermini et al. 2008; Rollo et al. 2006). This is a strong argument that the DNA obtained was indeed from Ötzi and not modern contamination.

DNA analyses have also been undertaken on Ötzi's stomach contents (Rollo et al. 2002). Samples of the intestinal content were obtained from the large and small bowel, and both animal and plant remains were targeted using specific PCR primers. The researchers suggested that Ötzi had the remains of two meals in his intestinal tract. His last meal included red deer (*Cervus elaphus*) meat, which followed a meal of ibex (*Capra ibex*). They also identified DNA of cereals and dicots (which includes plants like plums, beans, and peas). Pollen DNA found in his bowels indicates that he was probably traveling through a subalpine coniferous forest just prior to his death (Rollo et al. 2002).

The most recent chapter in the Ötzi story, not surprisingly, involves the publication of his entire genome (Keller et al. 2012). This study reports that Ötzi is most closely related to modern day Sardinians, that he was lactose intolerant, and he had type O blood. He also probably had dark hair and a genetic predisposition to cardiovascular disease. Interestingly, the authors also report that they recovered DNA sequences belonging to the bacterium associated with Lyme disease, suggesting that Ötzi may have been suffering from the infectious disease.

Ancient Genomics—Whole Genomes Recovered from Hair Samples

As has been discussed earlier, the development of NGS is dramatically extending our knowledge of genetic variation in extant, as well as ancient human and other hominin populations. While numerous groups are now using SNP chip technology to scan human genomes for known SNPs, the 1,000 genomes project (The 1000 Genomes Consortium) is attempting to sequence 1,000 entire human genomes representing individuals with ancestry from five major geographic regions: Europe, East Asia, South Asia, West Africa, and the Americas. These data will no doubt provide important information regarding modern human variation and adaptation. However, if we want to understand the processes involved in creating that variation and better understand how and when humans adapted and dispersed, we need aDNA sequences and further archaeological research.

Ancient Paleo-Eskimo from Greenland

As discussed in chapter five, the draft genome sequence of a Neanderthal was recently published (Green et al. 2010). The earliest modern human genome to be published was that of a 4,000 year old Paleo-Eskimo (Rasmussen et al. 2010). The DNA was obtained from hair samples recovered in Greenland from permafrost deposits associated with the Saqqaq culture, part of the Arctic small tool tradition dating to approximately 4750–2500 BP. Little is known about this culture or the people associated with it (Rasmussen et al. 2010).

In 2008, a research team sequenced the entire mtDNA genome from a Saqqaq hair sample and determined that the individual belonged to mtDNA haplogroup D2a1 (Gilbert et al. 2008). This haplotype has been previously recorded in modern Aleut people from the Commander Islands, in the Bering Sea, and Siberian Sireniki Yuit. It has not, however, been identified in Neo-Eskimo, who are characterized by haplogroup D3 or other Native American populations, who carry the D1 markers. Thus,

Gilbert and colleagues argued that there were different Siberian origins for the Saqqaq and Aleut than for other Native American populations. Alternatively, they point out, it is also possible that the Saqqaq lineages have been lost in other Native American groups due to demographic events, such as random genetic drift. The analysis of the entire genome in 2010 provided evidence that the owner of the hair samples was indeed male and that his Y chromosome belonged to haplogroup Q1a, which is a common haplogroup found in Siberian and Native American groups. When the researchers compared the genomic data with other native North American and northern Asian indigenous populations, they found that the Saqqaq genome was most closely related to the Koryaks and Chukchis and not Inuit. This, they argue, indicates that the ancestral populations of the Saqqaq and the Inuit diverged in northern Asia before the Saqqaq migrated into the New World. In addition to the important ancient demographic history provided by the Saqqaq DNA, the researchers were able to determine several phenotypic characteristics of the individual from which the hair derived. He had blood group A+, dark, thick hair, but with an increased risk of baldness, shovel shaped incisors (typical of Asian and Asian-derived populations), and dry ear-wax. A combination of twelve SNPs related to metabolism and body mass index indicated that he also probably had the typical cold adapted body type (short and stocky with relatively short appendages, which are related to heat retention).

The results presented by Rasmussen et al. (2010) provide an ideal example of the power of NGS and palaeogenomics for illuminating numerous aspects of the past from the population level to the specifics of individual phenotype. This is particularly so when samples can be recovered in ideal conditions. The Saqqaq sample was a hair sample, and hair provides particularly good aDNA with reduced risk of contamination (Gilbert et al. 2006), and it was also found in permafrost, which would increase the likelihood of recovery of good-quality DNA.

100-Year-Old Australian Aboriginal

In 2011, Rasmussen and colleagues were able to obtain a whole-genome sequence from a 100-year-old lock of hair that belonged to an Aboriginal Australian from western Australia (Rasmussen et al. 2011). The hair sample had been collected by the anthropologist, Dr. A. C. Haddon, from a young man living in Golden Ridge, near Kalgoorlie. In 1923, it was deposited into the Duckworth Laboratory Collections in the Museum of Archaeology and Ethnology at Cambridge University (Rasmussen et al. 2011).

The DNA analysis indicated that the young man who provided the hair sample had a Y chromosome that belonged to haplogroup

K-M526* and mtDNA belonging to a new subgroup of haplogroup O, which the researchers designated as haplogroup O1a. Both of these results are consistent with common Y chromosome and mtDNA lineages present in contemporary aboriginal Australians. Perhaps the most interesting results, however, were those obtained when the Australian data were compared to whole-genome studies of Africans, Europeans, and Asians. The results clearly indicated that Aboriginal Australians are the direct descendants of a population that was part of an early human dispersal event out of Africa or the Middle East into eastern Asia, dated to sometime between 62 and 75 KYA, significantly before other populations moved into Eurasia. The Australian genome clustered most closely with other Sahul samples from the highlands of New Guinea, and the data suggest that these groups had been genetically isolated from other populations for a period of between 15 and 30 KYA. Outside of Sahul, the closest populations to the Australians were the Munda speakers in India and the Aeta from the Philippines—both considered to be ancient "relics" of early out-of-Africa migrations. The authors argued that their data more strongly supported a multiple dispersal model into Asia, whereby the ancestors of Australian Aboriginal and related populations were part of the initial dispersal into Asia that split from the ancestors of other Eurasian populations. This split occurred prior to the split between Europeans and Asians. These later migrants into eastern Asia, arriving between 25 and 38 KYA, either replaced or strongly overrode the genetic signature of the first migration in most places in East Asia, other than in Sahul, and in some small isolated populations, such as among the Aeta (Rasmussen et al. 2011).

These most recent studies in ancient genomics highlight the potential for this type of aDNA research to provide information regarding phenotypic, as well as genotypic traits of ancient individuals. They also underscore the ability of aDNA to allow for the testing of various demographic models of population origin and dispersal. However, unless the samples are made available for scientists to analyze, the studies will never happen. It was critical that in the case of the sequencing of the Australian Aboriginal genome, the descendant community, the Goldfields Land and Sea Council provided full approval and support for the project (Rasmussen et al. 2011). As discussed in chapter four, this kind of support and approval should be obtained before research is undertaken and should be made publically available upon publication of results. It is hoped that the descendant communities who allow for the analysis of DNA belonging to their ancestors will find as much value in the data and results obtained as those of us who are interested in using genetic data for reconstructing the past.

Chapter Nine

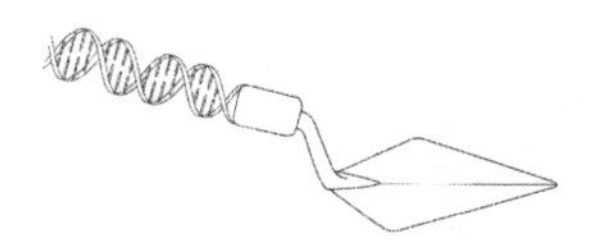

Conclusions

Biological anthropologists, or anyone else interested in the subject of archaeology, can pick up one of any number of books on the subject, including *Archaeology for Dummies* (White 2008). Granted, they could not, after reading the book, go out and undertake an archaeological excavation. They could, however, grasp key concepts such as stratigraphy, dating, and practical methods, such as why archaeologists dig square holes. They would even be introduced to some of the basic theoretical perspectives in the field. To our knowledge, there is no equivalent book, written for nonspecialists, on DNA applications in Anthropology. Our goal in writing this book was to provide such a primer—aimed primarily at archaeologists—so that they can be critical consumers of studies using DNA to address questions about human history. In addition to providing information on some of the basics of DNA, methods of analysis, processes, and issues that must be considered in both modern and aDNA studies, we have also, we hope, provided an overview of the various applications of DNA that might excite archaeologists and inspire more collaboration with DNA researchers.

One of the questions we are often asked by our archaeological colleagues is "how much will it cost us for you to 'do' the DNA of sample X for us?" As we mentioned in chapter one, anyone can go online and purchase a kit for obtaining information about their own DNA sequences simply by providing a cheek swab and credit card details. There are also labs that offer aDNA services, and our unofficial canvas of the cost of these services suggests it is about two to three times the cost of obtaining a radiocarbon date. It is not cheap, and there is no guarantee that there will be any DNA in the sample that you send. More importantly, what information is going to be returned? What genetic markers are they screening for? Are those markers appropriate for answering the actual question you are interested in?

For many reasons, obtaining a DNA sequence is not the same kind of thing as obtaining a radiocarbon date. Granted, a radiocarbon date needs to be calibrated, and thus the context of the sample is important. Still, a radiocarbon date provides some valuable information on its own—it tells you about the age of the material that was dated. A DNA sequence, on the other hand, tells you very little on its own. In order to obtain any kind of meaningful information, aDNA research in archaeology needs to be planned for from the beginning of the project. It should be incorporated in the field methodologies and the budget; there are ethical issues to consider; and there is, of course, the dreaded issue of contamination. Once the sample is recovered, it is critical to decide which part of the genome the analyst should target, and this requires specific knowledge regarding the questions being addressed. Once the sequence is generated, there is the matter of interpretation, which again, ideally requires specialist knowledge about DNA technology and of the anthropological and archaeological contexts. Thus, once again, we strongly recommend collaborations between archaeologists and DNA researchers. We particularly advocate collaborations with molecular anthropologists, who, unlike more generic biologists or geneticists, are hopefully engaged with the same theoretical perspectives and, ideally, the specific regional or topical issues being addressed by the archaeologist.

We have made every attempt in writing this book to provide the most up-to-date information on DNA technologies and their applications in addressing anthropological questions. However, we must reiterate that this is a rapidly changing field. New techniques and technologies are constantly evolving, and these often open up new opportunities for providing answers to questions we never even imagined we could address. In the last five years, for example, the Neanderthal genome has been sequenced (Green et al. 2010), providing information about aspects of their physical characteristics that are not preserved in the archaeological record. We can see how humans selected for particular coat colors in the process of domesticating horses (Ludwig et al. 2009). DNA has also been obtained from artifacts, such as feather cloaks, which can tell us about trade and exchange patterns at particular points in time (Hartnup et al. 2011). We can now assess the biological relatedness of ancient cemetery populations providing information regarding social structure, status, and other relationships that might not otherwise be apparent (Bouwman et al. 2008). Many of these types of studies were previously the domain of Hollywood producers and science fiction writers; and while we would not anticipate any real Jurassic Parks being established in the near future, we could not have anticipated these recent developments ten years ago.

Given that genetic technology changes so quickly, it is unrealistic to expect archaeologists will be able to stay up-to-date with the best molecular approaches to questions regarding prehistory. Therefore, communication needs to be ongoing between archaeologists and molecular anthropologists or other practitioners in DNA research. One forum for communication is the academic conference. In our experience, most DNA focused papers at the major archaeology conferences are clustered together in specific aDNA sessions. The focus is primarily on the methodology rather than the questions that are being addressed. We would like to encourage the integration of DNA-based studies in more regional or topic-focused sessions. This will allow for two-way communication. Not only do the archaeologists need to know about the techniques and results of DNA research, many DNA researchers would benefit from engaging with the broader subject matter and specific complexities of the anthropological issues being addressed.

One of the great advantages of the speed of development in the genetics and genomics fields is that efficiencies in speed in processing and cost of analyses will make DNA data more accessible and affordable. In the final days of writing this book, *New Scientist* announced the development of a sequencing machine called the "MinION." According to the article, "DNA sequencing can now be done on a device that plugs into your computer like a memory stick" (Graham-Rowe 2012:23). The manufacturer of a similar new sequencing machine claims that their "Ion Proton" will sequence an entire human genome in two hours for $1000 dollars (Graham-Rowe 2012:24). Despite these new advances, aDNA will always be more difficult, expensive, and time consuming than analyses of modern DNA due to the degraded nature of the DNA remaining in the samples—if any remains in the sample at all. Unfortunately, technological developments will never change that aspect of aDNA research.

Archaeologists and molecular anthropologists are natural allies in our attempts to understand the past; however, given different academic traditions, we tend to have limited shared academic training. Collaboration is logical, but it is something that needs to be worked at, on both sides. Archaeologists need to talk with geneticists and molecular anthropologists, and vice versa. First, however, they need to see the value of such collaboration. Then, they need to speak the same language and know what questions to ask. We hope that this volume has contributed to this cause.

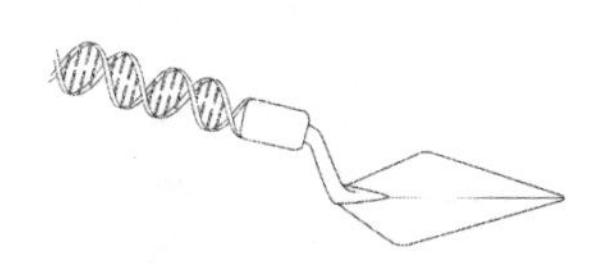

Notes

Chapter 2

1. There is some debate regarding the possibility of paternal leakage or paternal mtDNA becoming incorporated in the fertilized egg; although, it is generally believed that if this does happen, it is extremely rare and would not significantly affect the interpretations of mtDNA variation in human biogeographic studies—see Slate and Gemmell (2004) for discussion.
2. At the time of writing, Wikipedia provided a useful conversion table for many of the Y chromosome nomenclature systems commonly found in the literature: http://en.wikipedia.org/wiki/Conversion_table_for_Y_chromosome_haplogroups

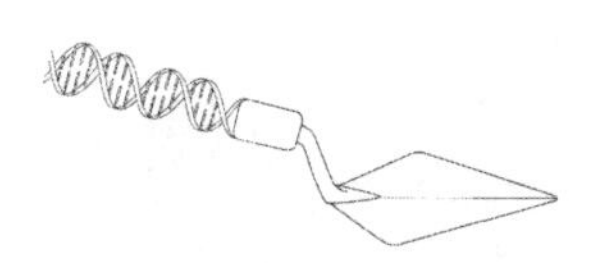

Glossary

Allele: One of two alternative forms of a gene or DNA sequence at a specific location on a chromosome.

Ascertainment bias: A systematic distortion in the measurement of the true frequency of a phenomenon, resulting from the way in which the data are collected or analyzed. Also known as a sampling bias.

Autosome: Any of the chromosomes in a cell that are not sex chromosomes. In humans, these are the chromosomes numbered one through twenty-two.

Bayesian methods: Statistical methods based on Bayes' theorem, which are used to calculate the conditional probability of an event given assumed prior knowledge. These estimates of probability may be updated based on new observations. Bayesian methods have been employed in many areas of genetics, particularly phylogenetics.

Bootstrapping: A resampling technique used to test the precision of sample statistics (e.g., medians, percentiles) of a given data set. This is achieved by random resampling and replacement of data points from the original data set. In phylogenetics, it gives a way of judging the strength of support for nodes on a phylogenetic tree.

Cambridge Reference Sequence: A mitochondrial DNA sequence published by researchers at Cambridge University in 1981, which has became the standard for describing changes to the mitochondrial sequences, particularly when doing genealogical research. As the original published sequence had a number of errors, a revised version of the sequence was published in 1999, which is known as the revised Cambridge Reference Sequence or rCRS. The rCRS is widely used for describing mtDNA variation.

Chromosome: An organized structure consisting of tightly coiled DNA and DNA-bound proteins found in the nuclei of cells. Human diploid cells (as opposed to gametes) normally have forty-six chromosomes.

Clade: A group of organisms that share a common ancestor; an evolutionary branch.

Codon: A group of three nucleotide bases that together form a unit of genetic code in a DNA or RNA molecule. A codon specifies which specific amino acid is inserted next in a polypeptide chain during protein synthesis.

Consensus tree: A tree that summarizes relationships from multiple phylogenetic trees.

Gametes: Cells involved in sex and reproduction, which contain half the number of chromosomes compared to most other cells, that is spermatozoa (male reproductive cells) and egg cells or ovum (female reproductive cells).

Gene: The fundamental physical and functional unit of heredity, which carries information from one generation to the next; a gene often consists of specific sequences of DNA nucleotides, which can code for proteins (i.e., codons). Individual genes may have variable forms (alleles) that form the basis of genetic variability.

Genome: The entirety of genetic material in a chromosome set (nuclear genome) or in the mitochondria (mitochondrial genome).

Germ cells: See Gametes.

Haplogroup: A group of similar haplotypes that share a common ancestor. In human mtDNA and Y chromosome phylogenies, haplogroups are usually assigned letters of the alphabet, and refinements consist of additional number and letter combinations. For instance, mtDNA haplogroup B4 (and subgroupings of this haplogroup) is common throughout Southeast Asia and B4a1a1a in the Pacific.

Haplotype: A set of linked genes or other genetic markers (e.g., mtDNA or Y chromosome mutations) that are generally inherited together as a unit.

Heteroplasmy: The presence of a mixture of more than one type of mitochondrial genome (or plastid genome in plants) within a cell or individual; this can be a result of "leakiness" from the male line where mitochondria from the tail of the spermatozoa are incorporated into the egg at the time of zygote formation, or alternatively from spontaneous novel mutations, which sometimes occur and which have not reached fixation in all cells of the body.

Hypervariable region: A portion of DNA characterized by multiple alleles for a single genetic locus; variability can manifest as accumulated base substitutions (as in the hypervariable regions 1 and 2, often referred to as HVR I and II, of the mitochondrial genome), or as nucleotide repeats (as in the case of nuclear DNA).

Maximum likelihood: A statistical phylogenetic method, wherein hypotheses about evolutionary history are evaluated in terms of the probability that the proposed model would give rise to the observed data set; the premise is that a history with a higher probability of reaching the state represented by the data set is more likely to be accurate than a history with a lower probability; thus the method searches for a tree with the highest likelihood.

Maximum parsimony: A simple, character-based, statistical method commonly used for estimating phylogenetic trees. A tree produced using this method assumes the least amount of evolutionary change between samples; based on the principle that simpler hypotheses are preferable to more complicated ones.

Meiosis: A type of cell division that gives rise to four gametes, each which possess half the chromosome number of the parent cell. This is the process by which diploid cells produce haploid gametes.

Mismatch distributions: A frequency graph of pairwise differences between DNA sequences in a sample. This tool can be used to infer the history of the population that gave rise to the sample of DNA sequences, for instance calculating the timing and degree of population expansion.

Mitochondrial DNA: A circular ring of DNA found in mitochondria. In mammals, mtDNA makes up less than 1% of the total cellular DNA. As mtDNA does not undergo recombination and is transmitted from mother to offspring, it has been deemed a useful matrilineal marker in molecular anthropological studies.

Mitosis: The process of somatic cell division by which a single cell divides to produce two identical daughter cells.

Molecular clock: The hypothesis that the systematic accumulation of mutations, or DNA sequence evolution, occurs at a relatively constant rate and thus, with an appropriate calibration point, can be used to determine the time at which two sequences or groups diverged.

Most recent common ancestor: The most recent individual from which all organisms in a group are directly descended—the deepest node on a phylogenetic tree from which all contemporary variants can trace their ancestry.

Mutation: A spontaneous and random change in the DNA sequence of a cell that can potentially cause it (and all cells derived from it) to differ in appearance or behaviour from the normal type. A mutation will only be passed from one generation to the next if it occurs in a sex cell.

Mutation rate: The rate at which a particular mutation occurs during some unit of time. Mutation rates differ between types of mutation, regions of the genome, and between species. Estimation of the mutation rate of neutral mutations is essential for the application of a molecular clock.

Neighbor-joining: A bottom-up clustering method used to create phylogenetic trees. These are usually based on data from DNA or protein sequences, and they require knowledge of the genetic distance between each pair of sequences in the tree.

Network methods: A method of visualizing the evolutionary relationship between samples (taxa) that recognizes reticulations or cycles. These may summarize or display incompatibilities of several phylogenetic trees given a particular data set.

Nonrecombining portion of the Y chromosome: DNA sequences in the Y chromosome that do not undergo recombination during meiosis. The sequences contain sex-determination genes and noncoding DNA. Accumulated mutations in these regions have been found to be useful in molecular anthropology for the purposes of detecting specific patrilineal lineages.

Nuclear DNA: The DNA contained within the nucleus of the cells of eukaryotic organisms that comprise portions inherited from two parents, one male and one female.

Nucleotides: The subunits that make up DNA and RNA. Organic compounds consisting of a nitrogen-containing purine or pyrimidine base-linked to a sugar (ribose or deoxyribose) and a phosphate group.

Outgroup: A group of organisms that serves as a reference group for determining the evolutionary relationship among a particular group of organisms that share a common ancestor. It is commonly used to root phylogenetic trees.

Parallel substitutions: Mutations that occur at the same site in the DNA sequence in independent lineages.

Parsimony: The principle that the most acceptable explanation of an event is the simplest, involving the fewest assumptions, or changes.

Phylogeny: A tree-like structure that represents the evolutionary history of an organism or group of related organisms.

Polymorphism (genetic): The occurrence of multiple variants of a characteristic (DNA sequences, allele, phenotypes, etc.) in a population. Generally a characteristic found at a frequency of above 1% is considered to be a polymorphic trait.

Pseudoautosomal region: Regions in the X and Y chromosomes. DNA sequences which allow them to pair up during meiotic cell division. This pairing up is essential to ensure the even distribution of X and Y chromosomes in the gametes of a male (i.e., one per cell). Genes within this region are not sex-linked.

Purines: A double-ringed organic nitrogenous base. Guanine.

Pyrimidines: A single-ringed organic nitrogenous base. Thymine and cytosine occur in nucleotides and nucleic acids.

Recombination: The rearrangement of genes that occurs when reproductive cells (gametes) are formed; results from the independent assortment of parental sets of chromosomes and exchange of chromosomal material that occurs during meiosis. This results in offspring that have a combination of characteristics different from that of their parents.

Reduction division: The pyrimidines in DNA. In RNA, the pyrimidines are uracil and cytosine.

Sex chromosomes: A chromosome that is involved in sex-determination of a species—in humans these are X and Y.

Silent Mutation: An alteration to the genetic code that has no apparent effect on the phenotype of an organism.

Somatic cells: All the cells in an animal or plant, except the reproductive cells.

Stem cells: An unspecialized cell that has the ability to go through numerous cycles of cell division while maintaining an undifferentiated state, and which has the capacity to give rise to any mature cell type via cell differentiation.

Zygote: A fertilized female gamete, the product of the fusion of the nucleus of an ovum with the nucleus of the sperm.

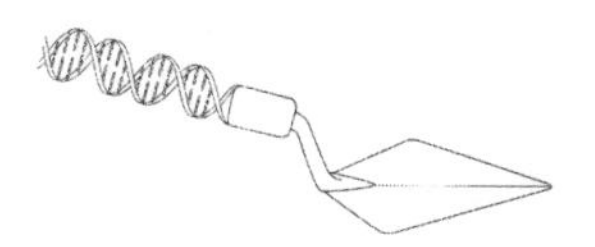

References

AAPA (American Association of Physical Anthropologists). 1996. "AAPA Statement on Biological Aspects of Race." *American Journal of Physical Anthropology* 101(4):569–570.

Abbey, D. M. 1999. "The Thomas Jefferson Paternity Case." *Nature* 397:32.

Abdulla, M. A., I. Ahmed, A. Assawamakin, J. Bhak, S. K. Brahmachari, G. C. Calacal, A. Chaurasia, C. H. Chen, J. M. Chen, Y. T. Chen, and others. 2009. "Mapping Human Genetic Diversity in Asia." *Science* 326(5959):1541–1545.

Achilli, A., S. Bonfiglio, A. Olivieri, A. Malusà, M. Pala, B. H. Kashani, U. A. Perego, P. Ajmone-Marsan, L. Liotta, O. Semino, and others. 2009. "The Multifaceted Origin of Taurine Cattle Reflected by the Mitochondrial Genome." *PLoS One* 4(6):1–7.

Achilli, A., A. Olivieri, P. Soares, H. Lancioni, B. H. Kashani, U. A. Perego, S. G. Nergadze, V. Carossa, M. Santagostino, S. Capomaccio, and others. 2012. "Mitochondrial Genomes from Modern Horses Reveal the Major Haplogroups that Underwent Domestication." *Proceedings of the National Academy of Sciences of the United States of America* 109(7):2449–2454.

Achilli, A., C. Rengo, V. Battaglia, M. Pala, A. Olivieri, S. Fornarino, C. Magri, R. Scozzari, N. Babudri, A. S. Santachiara-Benerecetti, and others. 2005. "Saami and Berbers—An Unexpected Mitochondrial DNA Link." *American Journal of Human Genetics* 76(5):883–886.

Adams, S. M., E. Bosch, P. L. Balaresque, S. J. Ballereau, A. C. Lee, E. Arroyo, A. M. López-Parra, M. Aler, M. S. G. Grifo, M. Brion, and others. 2008. "The Genetic Legacy of Religious Diversity and Intolerance: Paternal Lineages of Christians, Jews, and Muslims in the Iberian Peninsula." *American Journal of Human Genetics* 83(6):725–736.

Adcock, G. J., E. S. Dennis, S. Easteal, G. A. Huttley, L. S. Jermiin, W. J. Peacock, and A. Thorne. 2001. "Mitochondrial DNA Sequences in Ancient Australians: Implications for Modern Human Origins." *Proceedings of the National Academy of Sciences of the United States of America* 98(2):537–542.

Addison, D. J. and E. Matisoo-Smith. 2010. "Rethinking Polynesian Origins: A West-Polynesian Triple-I Model." *Archaeology in Oceania* 45(1):1–12.

Adler, C. J., W. Haak, D. Donlon, and A. Cooper. 2011. "Survival and Recovery of DNA from Ancient Teeth and Bones." *Journal of Archaeological Science* 38(5):956–964.

Allaby, R. G., T. A. Brown, and D. Q. Fuller. 2010. "A Simulation of the Effect of Inbreeding on Crop Domestication Genetics with Comments on the Integration of Archaeobotany and Genetics: A Reply to Honne and Heun." *Vegetation History and Archaeobotany* 19(2):151–158.

Allaby, R. G., D. Q. Fuller, and T. A. Brown. 2008. "The Genetic Expectations of a Protracted Model for the Origins of Domesticated Crops." *Proceedings of the National Academy of Sciences of the United States of America* 105(37):13982–13986.

Als, T. D., T. H. Jorgensen, A. D. Børglum, P. A. Petersen, O. Mors, and A. G. Wang. 2006. "Highly Discrepant Proportions of Female and Male Scandinavian and British Isles Ancestry within the Isolated Population of the Faroe Islands." *European Journal of Human Genetics* 14(4):497–504.

Anderson, S., A. T. Bankier, B. G. Barrell, M. H. L. de Bruijn, A. R. Coulson, J. Drouin, I. C. Eperon, D. P. Nierlich, B. A. Roe, F. Sanger, and others. 1981. "Sequence and Organisation of the Human Mitochondrial Genome." *Nature* 290:457–465.

Andrews, R. M., I. Kubacka, P. F. Chinnery, R. N. Lightowlers, D. M. Turnbull, and N. Howell. 1999. "Reanalysis and Revision of the Cambridge Reference Sequence for Human Mitochondrial DNA." *Nature Genetics* 23(2):147.

Armitage, S. J., S. A. Jasim, A. E. Marks, A. G. Parker, V. I. Usik, and H-P Uerpmann. 2011. "The Southern Route 'Out of Africa': Evidence for an Early Expansion of Modern Humans into Arabia." *Science* 331(6016):453–456.

Arnay-de-la-Rosa, M., E. Gonzalez-Reimers, R. Fregel, J. Velasco-Vazquez, T. Delgado-Darias, A. M. Gonzalez, and J. M. Larruga. 2007. "Canary Islands Aborigin Sex Determination Based on Mandible Parameters Contrasted by Amelogenin Analysis." *Journal of Archaeological Science* 34(9):1515–1522.

Arndt, A., W. Van Neer, B. Hellemans, J. Robben, F. Volckaert, and M. Waelkens. 2003. "Roman Trade Relationships at Sagalassos (Turkey) Elucidated by Ancient DNA of Fish Remains." *Journal of Archaeological Science* 30(9):1095–1105.

Arora, J., U. B. Singh, N. Suresh, T. Rana, C. Porwal, A. Kaushik, and J. N. Pande. 2009. "Characterization of Predominant *Mycobacterium tuberculosis* Strains from Different Subpopulations of India." *Infection Genetics and Evolution* 9(5):832–839.

Arroyo-García, R., L. Ruiz-García, L. Bolling, R. Ocete, M. A. López, C. Arnold, A. Ergul, G. Söylemezoglu, H. I. Uzun, F. Cabello, and others. 2006. "Multiple Origins of Cultivated Grapevine (*Vitis vinifera* L. ssp. *sativa*) Based on Chloroplast DNA Polymorphisms." *Molecular Ecology* 15(12):3707–3714.

Atkins, C., L. Reuffel, J. Roddy, M. Platts, H. Robinson, and R. Ward. 1988. "Rheumatic Disease in the Nuu-chah-nulth Native Indians of the Pacific Northwest." *Journal of Rheumatology* 15(4):684–690.

Atkinson, Q. D. 2011. Phonemic Diversity Supports a Serial Founder Effect Model of Language Expansion from Africa. *Science* 332(6027):346–349.

Atkinson, Q. D., R. D. Gray, and A. J. Drummond. 2008. "MtDNA Variation Predicts Population Size in Humans and Reveals a Major Southern Asian Chapter in Human Prehistory." *Molecular Biology and Evolution* 25(2):468–474.

Atkinson, Q. D., R. D. Gray, and A. J. Drummond. 2009. "Bayesian Coalescent Inference of Major Human Mitochondrial DNA Haplogroup Expansions in Africa." *Proceedings of the Royal Society B: Biological Sciences* 276(1655):367–373.

Aufderheide, A. C., W. Salo, M. Madden, J. Strietz, J. E. Buikstra, F. Guhl, B. Arriaza, C. Renier, L. E. Wittmers, G. Fornaciari, and others. 2004. "A 9,000-Year Record of Chagas' Disease." *Proceedings of the National Academy of Sciences* 101:2034–2039.

Baar, C., M. d'Abbadie, A. Vaisman, M. E. Arana, M. Hofreiter, R. Woodgate, T. A. Kunkel, and P. Holliger. 2011. "Molecular Breeding of Polymerases for Resistance to Environmental Inhibitors." *Nucleic Acids Research* 39(8):e51.

Balaresque, P., G. R. Bowden, S. M. Adams, H. Y. Leung, T. E. King, Z. H. Rosser, J. Goodwin, J-P. Moisan, C. Richard, A. Millward, and others. 2010. "A Predomi-nantly Neolithic Origin for European Paternal Lineages." *PLoS Biology* 8(1):e1000285.

Bandelt, H. J. 2005. "Mosaics of Ancient Mitochondrial DNA: Positive Indicators of Nonauthenticity." *European Journal of Human Genetics* 13(10):1106–1112.

Barnes, I. and M. G. Thomas. 2006. "Evaluating Bacterial Pathogen DNA Preservation in Museum Osteological Collections." *Proceedings of the Royal Society B: Biological Sciences* 273(1587):645–653.

Barnes, I., J. P. W. Young, and K. M. Dobney. 2000. "DNA-based Identification of Goose Species from Two Archaeological Sites in Lincolnshire." *Journal of Archaeological Science* 27(2):91–100.

Barnett, R., B. Shapiro, I. Barnes, S. Y. Ho, J. Burger, N. Yamaguchi, T. F. Higham, H. T. Wheeler, W. Rosendahl, A. V. Sher, and others. 2009. "Phylogeography of Lions (*Panthera leo* ssp.) Reveals Three Distinct Taxa and a Late Pleistocene Reduction in Genetic Diversity." *Molecular Ecology* 18(8):1668–1677.

Baron, H., S. Hummel, and B. Herrmann. 1996. "*Mycobacterium tuberculosis* Complex DNA in Ancient Human Bones." *Journal of Archaeological Science* 23(5):667–671.

Basu, A., N. Mukherjee, S. Roy, S. Sengupta, S. Banerjee, M. Chakraborty, B. Dey, M. Roy, B. Roy, N. Bhattacharyya, and others. 2003. "Ethnic India: A Genomic View, with Special Reference to Peopling and Structure." *Genome Research* 13(10):2277–2290.

Bathurst, R. R. and J. L. Barta. 2004. "Molecular Evidence of Tuberculosis Induced Hypertrophic Osteopathy in a 16th-century Iroquoian Dog." *Journal of Archaeological Science* 31(7):917–925.

Batini, C., V. Coia, C. Battaggia, J. Rocha, M. M. Pilkington, G. Spedini, D. Comas, G. Destro-Bisol, and F. Calafell. 2007. "Phylogeography of the Human Mitochondrial L1c Haplogroup: Genetic Signatures of the Prehistory of Central Africa." *Molecular Phylogenetics and Evolution* 43(2):635–644.

Batini, C., G. Ferri, G. Destro-Bisol, F. Brisighelli, D. Luiselli, P. Sánchez-Diz, J. Rocha, T. Simonson, A. Brehm, V. Montano, and others. 2011. "Signatures of the Preagricultural Peopling Processes in sub-Saharan Africa as Revealed by the Phylogeography of Early Y Chromosome Lineages." *Molecular Biology and Evolution* 28(9):2603–2613.

Batini, C., J. Lopes, D. M. Behar, F. Calafell, L. B. Jorde, L. van der Veen, L. Quintana-Murci, G. Spedini, G. Destro-Bisol, and D. Comas. 2011. "Insights into the Demographic History of African Pygmies from Complete Mitochondrial Genomes." *Molecular Biology and Evolution* 28(2):1099–1110.

Baudouin, L. and P. Lebrun. 2009. "Coconut (*Cocos nucifera* L.) DNA Studies Support the Hypothesis of an Ancient Austronesian Migration from Southeast Asia to America." *Genetic Resources and Crop Evolution* 56(2):257–262.

Behar, D. M., R. Villems, H. Soodyall, J. Blue-Smith, L. Pereira, E. Metspalu, R. Scozzari, H. Makkan, S. Tzur, D. Comas, and others. 2008. "The Dawn of Human Matrilineal Diversity." *American Journal of Human Genetics* 82(5):1130–1140.

Beja-Pereira, A., D. Caramelli, C. Lalueza-Fox, C. Vernesi, N. Ferrand, A. Casoli, F. Goyache, L. J. Royo, S. Conti, M. Lari, and others. 2006. "The Origin of European Cattle: Evidence from Modern and Ancient DNA." *Proceedings of the National Academy of Sciences of the United States of America* 103(21):8113–8118.

Beja-Pereira, A., P. R. England, N. Ferrand, S. Jordan, A. O. Bakhiet, M. A. Abdalla, M. Mashkour, J. Jordana, P. Taberlet, and G. Luikart. 2004. "African Origins of the Domestic Donkey." *Science* 304(5678):1781.

Beja-Pereira, A., G. Luikart, P. R. England, D. G. Bradley, O. C. Jann, G. Bertorelle, A. T. Chamberlain, T. P. Nunes, S. Metodiev, N. Ferrand, and others. 2003. "Gene-culture Coevolution between Cattle Milk Protein Genes and Human Lactase Genes." *Nature Genetics* 35(4):311–313.

Beleza, S., L. Gusmão, A. Amorim, A. Carracedo, and A. Salas. 2005. "The Genetic Legacy of Western Bantu Migrations." *Human Genetics* 117(4):366–375.

Bennett, K. D. and L. Parducci. 2006. DNA from Pollen: Principles and Potential. *The Holocene* 16(8):1031–1034.

Berniell-Lee, G., F. Calafell, E. Bosch, E. Heyer, L. Sica, P. Mouguiama-Daouda, L. van der Veen, J-M. Hombert, L. Quintana-Murci, and D. Comas. 2009. "Genetic and Demographic Implications of the Bantu Expansion: Insights from Human Paternal Lineages." *Molecular Biology and Evolution* 26(7):1581–1589.

Bianucci, R., L. Rahalison, E. R. Massa, A. Peluso, E. Ferroglio, and M. Signoli. 2008. "Technical Note: A Rapid Diagnostic Test Detects Plague in Ancient Human Remains:

An Example of the Interaction between Archeological and Biological Approaches (Southeastern France, 16th–18th Centuries).” *American Journal of Physical Anthropology* 136(3):361–367.

Binladen, J., C. Wiuf, M. T. P. Gilbert, M. Bunce, R. Barnett, G. Larson, A. D. Greenwood, J. Haile, S. Y. W. Ho, A. J. Hansen, and others. 2006. “Assessing the Fidelity of Ancient DNA Sequences Amplified from Nuclear Genes.” *Genetics* 172(2):733–741.

Biondi, G. and O. Rickards. 2007. “Race: The Extinction of a Paradigm.” *Annals of Human Biology* 34(6):588–592.

Birky, C. W. 1995. “Uniparental Inheritance of Mitochondrial and Chloroplast Genes: Mechanisms and Evolution.” *Proceedings of the National Academy of Sciences of the United States of America* 92(25):11331–11338.

Blanco-Verea, A., J. C. Jaime, M. Brión, and A. Carracedo. 2010. “Y-chromosome Lineages in Native South American Population.” *Forensic Science International: Genetics* 4(3):187–193.

Boessenkool, S., J. J. Austin, T. H. Worthy, P. Scofield, A. Cooper, P. J. Seddon, and J. M. Waters. 2009. “Relict or Colonizer? Extinction and Range Expansion of Penguins in Southern New Zealand.” *Proceedings of the Royal Society B: Biological Sciences* 276(1658):815–821.

Bogdanowicz, W., M. Allen, W. Branicki, M. Lembring, M. Gajewska, and T. Kupiec. 2009. “Genetic Identification of Putative Remains of the Famous Astronomer Nicolaus Copernicus.” *Proceedings of the National Academy of Sciences of the United States of America* 106(30):12279–12282.

Bökönyi, S. 1977. “The Introduction of Sheep Breeding to Europe.” *Ethnozootechnie* 21:65–70.

Bökönyi, S. 1997. “Zebus and Indian Wild Cattle.” *Anthropozoologica* 25/26:647–654.

Bollongino, R., J. Elsner, J. D. Vigne, and J. Burger. 2008. “Y-SNPs do not Indicate Hybridisation between European Aurochs and Domestic Cattle.” *PLoS One* 3(10):e3418.

Borson, N., F. Berdan, E. Stark, J. States, and P. J. Wettstein. 1998. “Origins of an Anasazi Scarlet Macaw Feather Artifact.” *American Antiquity* 63(1):131–142.

Bouwman, A. S. and T. A. Brown. 2005. “The Limits of Biomolecular Palaeopathology: Ancient DNA Cannot be Used to Study Venereal Syphilis.” *Journal of Archaeological Science* 32(5):703–713.

Bouwman, A. S., K. A. Brown, A. J. N. W. Prag, and T. A. Brown. 2008. “Kinship between Burials from Grave Circle B at Mycenae Revealed by Ancient DNA Typing.” *Journal of Archaeological Science* 35(9):2580–2584.

Bowcock, A. and L. Cavalli-Sforza. 1991. “The Study of Variation in the Human Genome.” *Genomics* 11(2):491–498.

Bower, M. A., M. Spencer, S. Matsumura, R. E. R. Nisbet, and C. J. Howe. 2005. “How Many Clones Need to be Sequenced from a Single Forensic or Ancient DNA Sample in Order to Determine a Reliable Consensus Sequence?” *Nucleic Acids Research* 33(8):2549–2556.

Bowler, J. M., H. Johnston, J. M. Olley, J. R. Prescott, R. G. Roberts, W. Shawcross, and N. A. Spooner. 2003. “New Ages for Human Occupation and Climatic Change at Lake Mungo, Australia.” *Nature* 421(6925):837–840.

Boyko, A. R., R. H. Boyko, C. M. Boyko, H. G. Parker, M. Castelhano, L. Corey, J. D. Degenhardt, A. Auton, M. Hedimbi, R. Kityo, and others. 2009. “Complex Population Structure in African Village Dogs and its Implications for Inferring Dog Domestication History.” *Proceedings of the National Academy of Sciences of the United States of America* 106(33):13903–13908.

Brace, L. 2005. *“Race” is a Four-letter Word: The Genesis of the Concept.* New York: Oxford University Press.

Bradley, B. J. 2008. “Reconstructing Phylogenies and Phenotypes: A Molecular View of Human Evolution.” *Journal of Anatomy* 212:337–353.

Bradley, D. G., D. E. MacHugh, P. P. Cunningham, and R. T. Loftus. 1996. "Mitochondrial Diversity and the Origins of African and European Cattle." *Proceedings of the National Academy of Sciences of the United States of America* 93(10):5131–5135.

Bradley, D. G., D. E. MacHugh, R. T. Loftus, R. S. Sow, C. H. Hoste, and E. P. Cunningham. 1994. "Zebu-taurine Variation in Y-chromosomal DNA: A Sensitive Assay for Genetic Introgression in West African Trypanotolerant Cattle Populations." *Animal Genetics* 25(S2):7–12.

Bramanti, B., M. G. Thomas, W. Haak, M. Unterlaender, P. Jores, K. Tambets, I. Antanaitis-Jacobs, M. N. Haidle, R. Jankauskas, C. J. Kind, and others. 2009. "Genetic Discontinuity between Local Hunter-Gatherers and Central Europe's First Farmers." *Science* 326(5949):137–140.

Breton, C., J. F. Terral, C. Pinatel, F. Médail, F. Bonhomme, and A. Bervillé. 2009. "The Origins of the Domestication of the Olive Tree." *Comptes Rendus Biologies* 332(12):1059–1064.

Briggs, A. W., J. M. Good, R. E. Green, J. Krause, T. Maricic, U. Stenzel, C. Lalueza-Fox, P. Rudan, D. Brajković, Z. Kućan, and others. 2009. "Targeted Retrieval and Analysis of Five Neandertal mtDNA Genomes." *Science* 325(5938):318–321.

Briggs, A. W., U. Stenzel, P. L. F. Johnson, R. E. Green, J. Kelso, K. Prüfer, M. Meyer, J. Krause, M. T. Ronan, M. Lachmann, and others. 2007. "Patterns of Damage in Genomic DNA Sequences from a Neandertal." *Proceedings of the National Academy of Sciences of the United States of America* 104(37):14616–14621.

Bromham, L. and D. Penny. 2003. "The Modern Molecular Clock." *Nature Reviews Genetics* 4:216–224.

Brotherton, P., P. Endicott, J. J. Sanchez, M. Beaumont, R. Barnett, J. Austin, and A. Cooper. 2007. "Novel High-resolution Characterization of Ancient DNA Reveals C > U-type Base Modification Events as the Sole Cause of Post Mortem Miscoding Lesions." *Nucleic Acids Research* 35(17):5717–5728.

Brown, K. S., C. W. Marean, A. I. R. Herries, Z. Jacobs, C. Tribolo, D. Braun, D. L. Roberts, M. C. Meyer, and J. Bernatchez. 2009. "Fire as an Engineering Tool of Early Modern Humans." *Science* 325(5942):859–862.

Brown, P., T. Sutikna, M. J. Morwood, R. P. Soejono, Jatmiko, E. W. Saptomo, and R. A. Due. 2004. "A New Small-bodied Hominin from the Late Pleistocene of Flores, Indonesia." *Nature* 431(7012):1055–1061.

Brown, T. A., M. K. Jones, W. Powell, and R. G. Allaby. 2009. "The Complex Origins of Domesticated Crops in the Fertile Crescent." *Trends in Ecology & Evolution* 24(2):103–109.

Brown, W. M., M. J. George, and A. C. Wilson. 1979. "Rapid Evolution of Animal Mitochondrial DNA." *Proceedings of the National Academy of Sciences of the United States of America* 76(4):1967–1971.

Bruford, M. W. and S. J. Townsend. 2006. "Mitochondrial DNA Diversity in Modern Sheep." In: M. A. Zeder, D. G. Bradley, E. Emshwiller, and B. D. Smith, editors. *Documenting Domestication: New Genetic and Archaeological Paradigms*. Berkeley: University of California Press. pp. 306–316.

Brumm, A., F. Aziz, G. D. van den Bergh, M. J. Morwood, M. K. Moore, I. Kurniawan, D. R. Hobbs, and R. Fullagar. 2006. "Early Stone Technology on Flores and its Implications for *Homo floresiensis*." *Nature* 441(7093):624–628.

Bunce, M., T. H. Worthy, M. J. Phillips, R. N. Holdaway, E. Willerslev, J. Haile, B. Shapiro, R. P. Scofield, A. Drummond, P. J. Kamp, and others. 2009. "The Evolutionary History of the Extinct Ratite Moa and New Zealand Neogene Paleogeography." *Proceedings of the National Academy of Sciences of the United States of America* 106(49):20646–20651.

Bunch, T. D., R. M. Michell, and A. Maciulis. 1990. "G-banded Chromosomes of the Gansu argali (*Ovis ammon jubata*) and their Implication on the Evolution of the Ovis karyotype." *Journal of Heredity* 81(3):227–230.

Burger, J., M. Kirchner, B. Bramanti, W. Haak, and M. G. Thomas. 2007. "Absence of the Lactase-persistence-associated Allele in Early Neolithic Europeans." *Proceedings of the National Academy of Sciences of the United States of America* 104(10):3736–3741.

Burney, D. A. and T. F. Flannery. 2005. "Fifty Millennia of Catastrophic Extinctions After Human Contact." *Trends in Ecology and Evolution* 20(7):395–401.

Cai, D. W., Z. W. Tang, L. Han, C. F. Speller, D. Y. Y. Yang, X. L. Ma, J. E. Cao, H. Zhu, and H. Zhou. 2009. "Ancient DNA Provides New Insights into the Origin of the Chinese Domestic Horse." *Journal of Archaeological Science* 36(3):835–842.

Calverie, J-M. and C. Notredame. 2006. *Bioinformatics for Dummies*. Hoboken, NJ: John Wiley and Sons.

Campbell, M. C. and S. A. Tishkoff. 2010. "The Evolution of Human Genetic and Phenotypic Variation in Africa." *Current Biology* 20(4):R166–R173.

Cann, R. L. 1993. "Obituary: A. C. Wilson 1935–1991." *Human Biology* 65(3):343–358.

Cann, R.L. 1994. "Ancient DNA—Herrmann, B., Hummel, S." *American Journal of Human Biology* 6(6):791–792.

Cann, R. L., M. Stoneking, and A. C. Wilson. 1987. "Mitochondrial-DNA and Human Evolution." *Nature* 325(6099):31–36.

Cano, R. J. and M. Borucki. 1995. "Revival and Identification of Bacterial Spores in 25- to 40-million-year-old Dominican Amber." *Science* 268(5213):1060–1064.

Cano, R. J., H. N. Poinar, N. J. Pieniazek, A. Acra, and G. O. Poinar. 1993. "Amplification and Sequencing of DNA from a 120–135-million-year-old Weevil." *Nature* 363(6429):536–538.

Cappellini, E., B. Chiarelli, L. Sineo, A. Casoli, A. Di Gioia, C. Vernesi, M. C. Biella, and D. Caramelli. 2004. "Biomolecular Study of the Human Remains from Tomb 5859 in the Etruscan Necropolis of Monterozzi, Tarquinia (Viterbo, Italy)." *Journal of Archaeological Science* 31(5):603–612.

Caramelli, D., C. Lalueza-Fox, S. Condemi, L. Longo, L. Milani, A. Manfredini, M. de Saint Pierre, F. Adoni, M. Lari, P. Giunti, and others. 2006. "A Highly Divergent mtDNA Sequence in a Neandertal Individual from Italy." *Current Biology* 16(16):R630–R632.

Carreel, F., D. G. de Leon, P. Lagoda, C. Lanaud, C. Jenny, J. P. Horry, and H. T. du Montcel. 2002. "Ascertaining Maternal and Paternal Lineage within *Musa* by Chloroplast and Mitochondrial DNA RFLP Analyses." *Genome* 45(4):679–692.

Cavalli-Sforza, L. L. 2005. "The Human Genome Diversity Project: Past, Present and Future." *Nature Reviews Genetics* 6(4):333–340.

Cavalli-Sforza, L. L., A. C. Wilson, C. R. Cantor, R. M. Cook-Deegan, and M. C. King. 1991. "Call for a Worldwide Survey of Human Genetic Diversity: A Vanishing Opportunity for the Human Genome Project." *Genomics* 11(2):490–491.

Chaubey, G., M. Metspalu, Y. Choi, R. Maegi, I. G. Romero, P. Soares, M. van Oven, D. M. Behar, S. Rootsi, G. Hudjashov, and others. 2011. "Population Genetic Structure in Indian Austroasiatic Speakers: The Role of Landscape Barriers and Sex-specific Admixture." *Molecular Biology and Evolution* 28(2):1013–1024.

Chen, Y-S., A. Olckers, T. G. Schurr, A. M. Kogelnik, K. Huoponen, and D. C. Wallace. 2000. "MtDNA Variation in the South African Kung and Khwe—and Their Genetic Relationships to Other African Populations." *American Journal of Human Genetics* 66(4):1362–1383.

Cheng, B., W. Tang, L. He, Y. Dong, J. Lu, Y. Lei, H. Yu, J. Zhang, and C. Xiao. 2008. "Genetic Imprint of the Mongol: Signal from Phylogeographic Analysis of Mitochondrial DNA." *Journal of Human Genetics* 53(10):905–913.

Cherni, L., V. Fernandes, J. B. Pereira, M. D. Costa, A. Goios, S. Frigi, B. Yacoubi-Loueslati, M. B. Amor, A. Slama, A. Amorin, and others. 2009. "Post-last Glacial Maximum Expansion from Iberia to North Africa Revealed by Fine Characterization of mtDNA H Haplogroup in Tunisia." *American Journal of Physical Anthropology* 139(2):253–260.

Clarke, A. C., M. K. Burtenshaw, P. A. McLenachan, D. L. Erickson, and D. Penny. 2006. "Reconstructing the Origins and Dispersal of the Polynesian Bottle Gourd (*Lagenaria siceraria*)." *Molecular Biology and Evolution* 23(5):893–900.

Clutton-Brock, J. 1971. "The Primary Food Animals of the Jericho Tell from the Proto-Neolithic to the Byzantine Period." *Levant* 3:41–55.

Clutton-Brock, J. 1989. "Cattle in Ancient North Africa." In: J. Clutton-Brock, editor. *The Walking Larder: Patterns of Domestication, Pastoralism and Predation*. London: Unwin Hyman. pp. 200–206.

Clutton-Brock, J. and H. P. Uerpmann. 1974. "The Sheep of Early Jericho." *Journal of Archaeological Science* 1(3):261–274.

Coble, M. D., O. M. Loreille, M. J. Wadhams, S. M. Edson, K. Maynard, C. E. Meyer, H. Niederstätter, C. Berger, B. Berger, A. B. Falsetti, and others. 2009. "Mystery Solved: The Identification of the Two Missing Romanov Children Using DNA Analysis." *PLoS One* 4(3):e4838.

Coelho, M., F. Sequeira, D. Luiselli, S. Beleza, and J. Rocha. 2009. "On the Edge of Bantu Expansions: Y Chromosome and Lactase Persistence Genetic Variation in Southwestern Angola." *BMC Evolutionary Biology* 9:80.

Colgan, D. J. 2001. "Commentary on G. J. Adcock, et al., 2001, 'Mitochondrial DNA Sequences in Ancient Australians: Implications for Modern Human Origins'." *Archaeology in Oceania* 36:168–169.

Colledge, S. 1998. "Identifying Pre-domestication Cultivation Using Multivariate Analysis." In: A. Damania, J. Valkoun, G. Willcox, and C. Qualset, editors. *The Origins of Agriculture and Plant Domestication*. Alleppo, Syria: International Center for Agricultural Research in the Dry Areas (ICARDA). pp. 121–131.

Collins, M. J., K. E. H. Penkman, N. Rohland, B. Shapiro, R. C. Dobberstein, S. Ritz-Timme, and M. Hofreiter. 2009. "Is Amino Acid Racemization a Useful Tool for Screening for Ancient DNA in Bone?" *Proceedings of the Royal Society B: Biological Sciences* 276(1669):2971–2977.

Colson, I. B., M. B. Richards, J. F. Bailey, B. C. Sykes, and R. E. M. Hedges. 1997. "DNA Analysis of Seven Human Skeletons Excavated from the Terp of Wijnaldum." *Journal of Archaeological Science* 24(10):911–917.

Coop, G., K. Bullaughey, F. Luca, and M. Przeworski. 2008. "The Timing of Selection at the Human FOXP2 Gene." *Molecular Biology and Evolution* 25(7):1257–1259.

Cooper, A., A. Rambaut, V. Macaulay, E. Willerslev, A. J. Hansen, and C. Stringer. 2001. "Human Origins and Ancient Human DNA." *Science* 292(5522):1655–1656.

Cordaux, R., R. Aunger, G. Bentley, I. Nasidze, S. M. Sirajuddin, and M. Stoneking. 2004. "Independent Origins of Indian Caste and Tribal Paternal Lineages." *Current Biology* 14(3):231–235.

Correns, C. 1900. "G. Mendels Regel Über das Verhalten der Nachkommenschaft der Rassenbastarde." *Berichte der Deutschen Botanischen Gesellschaft* 18:158–168.

Crubézy, E., L. Legal, G. Fabas, H. Dabernat, and B. Ludes. 2006. "Pathogeny of Archaic Mycobacteria at the Emergence of Urban Life in Egypt (3400 BC)." *Infection Genetics and Evolution* 6(1):13–21.

Cruciani, F., R. La Fratta, B. Trombetta, P. Santolamazza, D. Sellitto, E. B. Colomb, J-M. Dugoujon, F. Crivellaro, T. Benincasa, R. Pascone, and others. 2007. "Tracing Past Human Male Movements in Northern/Eastern Africa and Western Eurasia: New Clues from Y-chromosomal Haplogroups E-M78 and J-M12." *Molecular Biology and Evolution* 24(6):1300–1311.

Cruciani, F., P. Santolamazza, P. Shen, V. Macaulay, P. Moral, A. Olckers, D. Modiano, S. Holmes, G. Destro-Bisol, V. Coica, and others. 2002. "A Back Migration from Asia to Sub-Saharan Africa is Supported by High Resolution Analysis of Human Y Chromosome Haplotypes." *American Journal of Human Genetics* 70(5):1197–1214.

Cruciani, F., B. Trombetta, A. Massaia, G. Destro-Bisol, D. Sellitto, and R. Scozzari. 2011. "A Revised Root for the Y Chromosomal Phylogenetic Tree: The Origin of Patrilineal Diversity in Africa." *American Journal of Human Genetics* 88(6):814–818.

Cruz, F., C. Vilà, and M. T. Webster. 2008. "The Legacy of Domestication: Accumulation of Deleterious Mutations in the Dog Genome." *Molecular Biology and Evolution* 25(11):2331–2336.

Cunha, E., M. L. Fily, I. Clisson, A. L. Santos, A. M. Silva, C. Umbelino, P. César, A. Corte-Real, E. Crubézy, and B. Ludes. 2000. "Children at the Convent: Comparing Historical Data, Morphology and DNA Extracted from Ancient Tissues for Sex Diagnosis at Santa Clara-a-Velha (Coimbra, Portugal)." *Journal of Archaeological Science* 27(10):949–952.

Darlu, P. and P. Tassy. 1987. "Disputed African Origin of Human Populations." *Nature* 329(6135):111.

Daskalaki, E., C. Anderung, L. Humphrey, and A. Götherström. 2011. "Further Developments in Molecular Sex Assignment: A Blind Test of 18th and 19th Century Human Skeletons." *Journal of Archaeological Science* 38(6):1326–1330.

de Filippo, C., C. Barbieri, M. Whitten, S. W. Mpoloka, E. D. Gunnarsdóttir, K. Bostoen, T. Nyambe, K. Beyer, H. Schreiber, P. de Knijff, and others. 2011. "Y-Chromosomal Variation in Sub-Saharan Africa: Insights into the History of Niger-Congo Groups." *Molecular Biology and Evolution* 28(3):1255–1269.

De La Cruz, I., A. González-Oliver, B. M. Kemp, J. A. Román, D. G. Smith, and A. Torre-Blanco. 2008. "Sex Identification of Children Sacrificed to the Ancient Aztec Rain Gods in Tlatelolco." *Current Anthropology* 49(3):519–526.

De Vries, H. 1900. "Sur la loi de disjonction des hybrides." *Comptes Rendus de l'Academie des Sciences* (Paris) 130:845–847.

Deguilloux, M-F., L. Bertel, A. Celant, M. H. Pemonge, L. Sadori, D. Magri, and R. J. Petit. 2006. "Genetic Analysis of Archaeological Wood Remains: First Results and Prospects." *Journal of Archaeological Science* 33: 1216–1227.

Deguilloux, M-F., J. Moquel, M. H. Pemonge, G. Colombeau. 2009. "Ancient DNA Supports Lineage Replacement in European Dog Gene Pool: Insight into Neolithic Southeast France." *Journal of Archaeological Science* 36:513–519.

Deguilloux, M-F., M-H. Pemonge, V. Dubut, S. Hughes, C. Hänni, L. Chollet, E. Conte, and P. Murail. 2011. "Human Ancient and Extant mtDNA from the Gambier Islands (French Polynesia): Evidence for an Early Melanesian Maternal Contribution and New Perspectives into the Settlement of Easternmost Polynesia." *American Journal of Physical Anthropology* 144(2):248–257.

Delcourt, P. A., H. R. Delcourt, P. A. Cridlebaugh, and J. Chapman. 1986. "Holocene Ethnobotanical and Paleoecological Record of Human Impact on Vegetation in the Little Tennessee River Valley, Tennessee." *Quaternary Research* 25(3):330–349.

Delfin, F., S. Myles, Y. Choi, D. Hughes, R. Illek, M. van Oven, B. Pakendorf, M. Kayser, and M. Stoneking. 2011. "Bridging Near and Remote Oceania: MtDNA and NRY Variation in the Solomon Islands." *Molecular Biology and Evolution* 29(2):545–564.

Denham, T. P., S. G. Haberle, C. Lentfer, R. Fullagar, J. Field, M. Therin, N. Porch, and B. Winsborough. 2003. "Origins of Agriculture at Kuk Swamp in the Highlands of New Guinea." *Science* 301(5630):189–193.

DeSalle, R., J. Gatesy, W. Wheeler, and D. Grimaldi. 1992. "DNA Sequences from a Fossil Termite in Oligo-Miocene Amber and Their Phylogenetic Implications." *Science* 257(5078):1933–1936.

Dillehay, T. D., C. Ramírez, M. Pino, M. B. Collins, J. Rossen, and J. D. Pino-Navarro. 2008. "Monte Verde: Seaweed, Food, Medicine, and the Peopling of South America." *Science* 320(5877):784–786.

Dissing, J., M. A. Kristinsdottir, and C. Friis. 2008. "On the Elimination of Extraneous DNA in Fossil Human Teeth with Hypochlorite." *Journal of Archaeological Science* 35(6):1445–1452.

Dizon, A., C. S. Baker, F. Cipriano, G. Lento, P. Palsbøll, and R. Reeves. (eds.) 2000. "Molecular Genetic Identification of Whales, Dolphins, and Porpoises." *Proceedings of a Workshop on the Forensic Use of Molecular Techniques to Identify Wildlife Products in the Marketplace*. La Jolla, CA, USA, 14–16 June 1999. U.S. Department of Commerce, NOAA Technical Memorandum, NOAA-TM-NMFS-SWFSC-2865.2. p. xi.

Dobberstein, R. C., J. Huppertz, N. von Wurmb-Schwark, and S. Ritz-Timme. 2008. "Degradation of Biomolecules in Artificially and Naturally Aged Teeth: Implications for Age Estimation Based on Aspartic Acid Racemization and DNA Analysis." *Forensic Science International* 179(2–3):181–191.

Dobzhansky, T. 1937. *Genetics and the Origin of Species*. New York: Columbia University Press.

Doebley, J. 2004. "The Genetics of Maize Evolution." *Annual Review of Genetics* 38:37–59.

Donoghue, H. D., I. Hershkovitz, D. E. Minnikin, G. S. Besra, O. Y. C. Lee, E. Galili, C. L. Greenblatt, E. Lemma, M. Spigelman, and G. K. Bar-Gal. 2009. "Biomolecular Archaeology of Ancient Tuberculosis: Response to 'Deficiencies and Challenges in the Study of Ancient Tuberculosis DNA' by Wilbur et al. (2009)." *Journal of Archaeological Science* 36(12):2797–2804.

Donoghue, H. D., A. Marcsik, C. Matheson, K. Vernon, E. Nuorala, J. E. Molto, C. L. Greenblatt, and M. Spigelman. 2005. "Co-infection of *Mycobacterium tuberculosis* and *Mycobacterium leprae* in Human Archaeological Samples: A Possible Explanation for the Historical Decline of Leprosy." *Proceedings of the Royal Society B: Biological Sciences* 272(1561):389–394.

Donoghue, H. D., M. Spigelman, C. L. Greenblatt, G. Lev-Maor, G. K. Bar-Gal, C. Matheson, K. Vernon, A. G. Nerlich, and A. R. Zink. 2004. "Tuberculosis: From Prehistory to Robert Koch, as Revealed by Ancient DNA." *Lancet Infectious Diseases* 4(9):584–592.

Donoghue, H. D., M. Spigelman, J. Zias, A. M. Gernaey-Child, and D. E. Minnikin. 1998. "*Mycobacterium tuberculosis* Complex DNA in Calcified Pleura from Remains 1400 Years Old." *Letters in Applied Microbiology* 27(5):265–269.

Drancourt, M., G. Aboudharam, M. Signoli, O. Dutour, and D. Raoult. 1998. "Detection of 400-year-old *Yersinia pestis* DNA in Human Dental Pulp: An Approach to the Diagnosis of Ancient Septicemia." *Proceedings of the National Academy of Sciences of the United States of America* 95(21):12637–12640.

Drancourt, M., M. Signoli, L. V. Dang, B. Bizot, V. Roux, S. Tzortzis, and D. Raoult. 2007. "*Yersinia pestis orientalis* in Remains of Ancient Plague Patients." *Emerging Infectious Diseases* 13(2):332–333.

Driscoll, C. A., M. Menotti-Raymond, A. L. Roca, K. Hupe, W. E. Johnson, E. Geffen, E. H. Harley, M. Delibes, D. Pontier, A. C. Kitchener, and others. 2007. "The Near Eastern Origin of Cat Domestication." *Science* 317(5837):519–523.

Drummond, A. J. and A. Rambaut. 2007. "BEAST: Bayesian Evolutionary Analysis by Sampling Trees." *BMC Evolutionary Biology* 7:214.

Duarte, C., J. Maurício, P. B. Pettitt, P. Souto, E. Trinkaus, H. van der Plicht, and J. Zilho. 1999. "The Early Upper Paleolithic Human Skeleton from the Abrigo do Lagar Velho (Portugal) and Modern Human Emergence in Iberia." *Proceedings of the National Academy of Sciences of the United States of America* 96(13):7604–7609.

Eckhardt, R. B., M. H. Wolpoff, and A. G. Thorne. 1993. "Multiregional Evolution." *Science* 262(5136):973–974.

Edwards, C. J., R. Bollongino, A. Scheu, A. Chamberlain, A. Tresset, J-D. Vigne, J. F. Baird, G. Larson, S. Y. W. Ho, T. H. Heupink, and others. 2007. "Mitochondrial DNA Analysis Shows a Near Eastern Neolithic Origin for Domestic Cattle and No Indication of Domestication of European Aurochs." *Proceedings of the Royal Society B: Biological Sciences* 274(1616):1377–1385.

Efron, B. and G. Gong. 1983. "A Leisurely Look at the Bootstrap, the Jackknife, and Cross–validation." *The American Statistician* 37(1):36–48.

Ehret, C. 2001. "Bantu Expansions: Re-envisioning a Central Problem of Early African History." *International Journal of African Historical Studies* 34(1):5–41.

Elbaum, R., C. Melamed-Bessudo, E. Boaretto, E. Galili, S. Lev-Yadun, A. A. Levy, and S. Weiner. 2006. "Ancient Olive DNA in Pits: Preservation, Amplification and Sequence Analysis." *Journal of Archaeological Science* 33(1):77–88.

Ellis, N., M. Hammer, M. E. Hurles, M. A. Jobling, T. Karafet, T. E. King, P. de Knijff, A. Pandya, A. Redd, F. R. Santos, and others. 2002. "A Nomenclature System for the Tree of Human Y-chromosomal Binary Haplogroups." *Genome Research* 12(2):339–348.

Enard, W., M. Przeworski, S. E. Fisher, C. S. L. Lai, V. Wiebe, T. Kitano, A. P. Monaco, and S. Pääbo. 2002. "Molecular Evolution of FOXP2, a Gene Involved in Speech and Language." *Nature* 418:869–872.

Enattah, N. S., T. G. K. Jensen, M. Nielsen, R. Lewinski, M. Kuokkanen, H. Rasinpera, H. El-Shanti, J. K. Seo, M. Alifrangis, I. F. Khalil, and others. 2008. "Independent Introduction of Two Lactase-persistence Alleles into Human Populations Reflects Different History of Adaptation to Milk Culture." *American Journal of Human Genetics* 82(1):57–72.

Endicott, P. and S. Y. W. Ho. 2008. "A Bayesian Evaluation of Human Mitochondrial Substitution Rates." *American Journal of Human Genetics* 82(4):895–902.

Endicott, P., J. J. Sanchez, E. Metspalu, D. M. Behar, and T. Kivisild. 2007. "The Unresolved Location of Ötzi's mtDNA within Haplogroup K." *American Journal of Physical Anthropology* 132:590–593.

Ennafaa, H., V. M. Cabrera, K. K. Abu-Amero, A. M. González, M. B. Amor, R. Bouhaha, N. Dzimiri, A. B. Elgaaïed, and J. M. Larruga. 2009. "Mitochondrial DNA Haplogroup H Structure in North Africa." *BMC Genetics* 10:8.

Eren, Metin, I. (ed.) 2012. *Hunter-gatherer Behavior, Human Response during the Younger Dryas*. Walnut Creek: Left Coast Press.

Erickson, D. L., B. D. Smith, A. C. Clarke, D. H. Sandweiss, and N. Tuross. 2005. "An Asian Origin for a 10,000-year-old Domesticated Plant in the Americas." *Proceedings of the National Academy of Sciences of the United States of America* 102(51):18315–18320.

Ermini, L., C. Olivieri, E. Rizzi, G. Corti, R. Bonnal, P. Soares, S. Luciani, I. Marota, G. De Bellis, M. B. Richards, and others. 2008. "Complete Mitochondrial Genome Sequence of the Tyrolean Iceman." *Current Biology* 18(21):1687–1693.

Faerman, M., G. K. Bar-Gal, D. Filon, C. L. Greenblatt, L. Stager, A. Oppenheim, and P. Smith. 1998. "Determining the Sex of Infanticide Victims from the Late Roman Era through Ancient DNA Analysis." *Journal of Archaeological Science* 25(9):861–865.

Faerman, M., D. Filon, G. Kahila, C. L. Greenblatt, P. Smith, and A. Oppenheim. 1995. "Sex Identification of Archaeological Human Remains Based on Amplification of the X and Y Amelogenin Alleles." *Gene* 167(1–2):327–332.

Faerman, M., G. Kahila, P. Smith, C. Greenblatt, L. Stager, D. Filon, and A. Oppenheim. 1997. "DNA Analysis Reveals the Sex of Infanticide Victims." *Nature* 385(6613):212–213.

Fagundes, N. J. R., R. Kanitz, R. Eckert, A. C. S. Valls, M. R. Bogo, F. M. Salzano, D. G. Smith, W. A. Silva, M. A. Zago, A. K. Ribeiro-dos-Santos, and others. 2008. "Mitochondrial Population Genomics Supports a Single Pre-Clovis Origin with a Coastal Route for the Peopling of the Americas." *American Journal of Human Genetics* 82(3):583–592.

Falush, D., T. Wirth, B. Linz, J. K. Pritchard, M. Stephens, M. Kidd, M. J. Blaser, D. Y. Graham, S. Vacher, G. I. Perez-Perez, and others. 2003. "Traces of Human Migrations in *Helicobacter pylori* Populations." *Science* 299(5612):1582–1585.

Felsenstein, J. 1981. "Evolutionary Trees from DNA Sequences—A Maximum-likelihood Approach." *Journal of Molecular Evolution* 17(6):368–376.

Felsenstein, J. 1993. *PHYLIP: Phylogeny Inference Package*. University of Washington. Seattle, WA.

Fernandes, A., A. M. Iñiguez, V. S. Lima, S. de Souza, L. F. Ferreira, A. C. P. Vicente, and A. M. Jansen. 2008. "Pre-Columbian Chagas Disease in Brazil: *Trypanosoma cruzi I* in the Archaeological Remains of a Human in Peruaçu Valley, Minas Gerais, Brazil." *Memórias Do Instituto Oswaldo Cruz* 103(5):514–516.

Fernandes, V., F. Alshamali, M. Alves, M. D. Costa, J. B. Pereira, N. M. Silva, L. Cherni, N. Harich, V. Cerny, P. Soares, and others. 2012. "The Arabian Cradle: Mitochondrial Relicts of the First Steps along the Southern Route Out of Africa." *American Journal of Human Genetics* 90(2):347–355.

Fernández, H., S. Hughes, J-D. Vigne, D. Helmer, G. Hodgins, C. Miquel, C. Ha¨nni, G. Luikart, and P. Taberlet. 2006. "Divergent mtDNA Lineages of Goats in an Early Neolithic Site, Far from the Initial Domestication Areas." *Proceedings of the National Academy of Sciences of the United States of America* 103(42):15375–15379.

Firestone, R. B., A. West, J. P. Kennett, L. Becker, T. E. Bunch, Z. S. Revay, P. H. Schultz, T. Belgya, D. J. Kennett, J. M. Erlandson, and others. 2007. "Evidence for an Extraterrestrial Impact 12,900 Years Ago that Contributed to the Megafaunal Extinctions and the Younger Dryas Cooling." *Proceedings of the National Academy of Sciences of the United States of America* 104(41):16016–16021.

Fitch, W. M. and E. Margoliash. 1967. "Construction of Phylogenetic Trees." *Science* 155:279–284.

Fitzpatrick, S. M. and R. Callaghan. 2009. "Examining Dispersal Mechanisms for the Translocation of Chicken (*Gallus gallus*) from Polynesia to South America." *Journal of Archaeological Science* 36(2):214–223.

Fletcher, H. A., H. D. Donoghue, J. Holton, I. Pap, and M. Spigelman. 2003. "Widespread Occurrence of *Mycobacterium tuberculosis* DNA from 18th–19th Century Hungarians." *American Journal of Physical Anthropology* 120(2):144–150.

Fletcher, H. A., H. D. Donoghue, G. M. Taylor, A. G. M. van der Zanden, and M. Spigelman. 2003. "Molecular Analysis of *Mycobacterium tuberculosis* DNA from a Family of 18th Century Hungarians." *Microbiology* 149:143–151.

Foster, E. A., M. A. Jobling, P. G. Taylor, P. Donnelly, P. de Knijff, R. Mieremet, T. Zerjal, and C. Tyler-Smith. 1998. "Jefferson Fathered Slave's Last Child." *Nature* 396(6706):27–28.

Foster, E. A., M. A. Jobling, P. G. Taylor, P. Donnelly, P. de Knijff, R. Mieremet, T. Zerjal, and C. Tyler-Smith. 1999. "The Thomas Jefferson Paternity Case—Reply." *Nature* 397(6714):32.

Francis, C. M., A. V. Borisenko, N. V. Ivanova, J. L. Eger, B. K. Lim, A. Guillén-Servent, S. V. Kruskop, I. Mackie, and P. D. N. Hebert. 2010. "The Role of DNA Barcodes in Understanding and Conservation of Mammal Diversity in Southeast Asia." *PLoS One* 5(9):e12575.

Franklin, R. E. and R. G. Gosling. 1953. "Molecular Configuration of Sodium Thymonucleate." *Nature* 4356:740–741.

Fraser, M., S. Sten, and A. Götherström. 2012. "Neolithic Hedgehogs (*Erinaceus europaeus*) from the Island of Gotland Show Early Contacts with the Swedish Mainland." *Journal of Archaeological Science* 39(2):229–233.

Fricker, E. J., M. Spigelman, and C. R. Fricker. 1997. "The Detection of *Escherichia coli* DNA in the Ancient Remains of Lindow Man Using Polymerase Chain Reaction." *Letters in Applied Microbiology* 24:351–354.

Friedlaender, J., T. Schurr, F. Gentz, G. Koki, F. Friedlaender, G. Horvat, P. Babb, S. Cerchio, F. Kaestle, M. Schanfield, and others. 2005. "Expanding Southwest Pacific Mitochondrial Haplogroups P and Q." *Molecular Biology and Evolution* 22(6):1506–1517.

Friedlaender, J. S., F. R. Friedlaender, J. A. Hodgson, M. Stoltz, G. Koki, G. Horvat, S. Zhadanov, T. G. Schurr, and D. A. Merriwether. 2007. "Melanesian mtDNA Complexity." *PLoS One* 2(2):e248.

Friedlaender, J. S., F. R. Friedlaender, F. A. Reed, K. K. Kidd, J. R. Kidd, G. K. Chambers, R. A. Lea, J. H. Loo, G. Koki, J. A. Hodgson, and others. 2008. "The Genetic Structure of Pacific Islanders." *PLoS Genetics* 4(1).

Frumkin, A., O. Bar-Yosef, and H. P. Schwarcz. 2011. "Possible Paleohydrologic and Paleoclimatic Effects on Hominin Migration and Occupation of the Levantine Middle Paleolithic." *Journal of Human Evolution* 60(4):437–451.

Gabriel, S. E., K. N. Brigman, B. H. Koller, R. C. Boucher, and M. J. Stutts. 1994. "Cystic-fibrosis Heterozygote Resistance to Cholera-toxin in the Cystic-Fibrosis Mouse Model." *Science* 266(5182):107–109.

Gaieski, J. B., A. C. Owings, M. G. Vilar, M. C. Dulik, D. F. Gaieski, R. M. Gittelman, J. Lindo, L. Gau, T. G. Schurr, and The Genographic Consortium. 2011. "Genetic Ancestry and Indigenous Heritage in a Native American Descendant Community in Bermuda." *American Journal of Physical Anthropology* 146(3):392–405.

Gamble, C., W. Davies, P. Pettitt, and M. Richards. 2004. "Climate Change and Evolving Human Diversity in Europe during the Last Glacial." *Philosophical Transactions of the Royal Society of London Series B: Biological Sciences* 359(1442):243–254.

Gao, X., P. Zimmet, and S. W. Serjeantson. 1992. "HLA-DR, DQ Sequence Polymorphisms in Polynesians, Micronesians, and Javanese." *Human Immunology* 34(3):153–161.

Garrigan, D., S. B. Kingan, M. M. Pilkington, J. A. Wilder, M. P. Cox, H. Soodyall, B. Strassmann, G. Destro-Bisol, P. de Knijff, A. Novelletto, and others. 2007. "Inferring Human Population Sizes, Divergence Times and Rates of Gene Flow from Mitochondrial, X and Y Chromosome Resequencing Data." *Genetics* 177(4):2195–2207.

Gautier, A. 1984. "Archaeozoology of the Bir Kiseiba Region, East Sahara." In: F. Wendorf and R. Schild, editors. *Cattle-keepers of the Eastern Sahara: The Neolithic of Bir Kiseiba*. Dallas: Southern Methodist University. pp. 49–72.

Geigl, E-M. 2008. "Palaeogenetics of Cattle Domestication: Methodological Challenges for the Study of Fossil Bones Preserved in the Domestication Centre in Southwest Asia." *Comptes Rendus Palevol* 7(2–3):99–112.

Geist, V. 1991. "On the Taxonomy of Giant Sheep (*Ovis ammon* Linnaeus, 1766)." *Canadian Journal of Zoology* 69(3):706–723.

Gerbault, P., C. Moret, M. Currat, and A. Sanchez-Mazas. 2009. "Impact of Selection and Demography on the Diffusion of Lactase Persistence." *PLoS One* 4(7):e6369.

Germonpré, M., M. V. Sablin, R. E. Stevens, R. E. M. Hedges, M. Hofreiter, M. Stiller, and V. R. Després. 2009. "Fossil Dogs and Wolves from Palaeolithic Sites in Belgium, the Ukraine and Russia: Osteometry, Ancient DNA and Stable Isotopes." *Journal of Archaeological Science* 36(2):473–490.

Gernaey, A. M., D. E. Minnikin, M. S. Copley, R. A. Dixon, J. C. Middleton, and C. A. Roberts. 2001. "Mycolic Acids and Ancient DNA Confirm an Osteological Diagnosis of Tuberculosis." *Tuberculosis* 81(4):259–265.

Gilbert, M. T. P., J. Cuccui, W. White, N. Lynnerup, R. W. Titball, A. Cooper, and M. B. Prentice. 2004. "Absence of *Yersinia pestis*-specific DNA in Human Teeth from Five European Excavations of Putative Plague Victims." *Microbiology* 150:341–354.

Gilbert, M. T. P., D. I. Drautz, A. M. Lesk, S. Y. W. Ho, J. Qi, A. Ratan, C. H. Hsu, A. Sher, L. Dalén, A. Götherström, and others. 2008. "Intraspecific Phylogenetic Analysis of Siberian Woolly Mammoths Using Complete Mitochondrial Genomes." *Proceedings of the National Academy of Sciences of the United States of America* 105(24):8327–8332.

Gilbert, M. T. P., A. J. Hansen, E. Willerslev, L. Rudbeck, I. Barnes, N. Lynnerup, and A. Cooper. 2003. "Characterization of Genetic Miscoding Lesions Caused by Postmortem Damage." *American Journal of Human Genetics* 72(1):48–61.

Gilbert, M. T. P., A. J. Hansen, E. Willerslev, G. Turner-Walker, and M. Collins. 2006. "Insights into the Processes Behind the Contamination of Degraded Human Teeth and Bone Samples with Exogenous Sources of DNA." *International Journal of Osteoarchaeology* 16(2):156–164.

Gilbert, M. T. P., D. L. Jenkins, A. Götherström, N. Naveran, J. J. Sanchez, M. Hofreiter, P. F. Thomsen, J. Binladen, T. F. G. Higham, R. M. Yohe, and others. 2008. "DNA from pre-Clovis Human Coprolites in Oregon, North America." *Science* 320(5877):786–789.

Gilbert, M. T. P., T. Kivisild, B. Grønnow, P. K. Andersen, E. Metspalu, M. Reidla, E. Tamm, E. Axelsson, A. Götherström, P. F. Campos, and others. 2008. "Paleo-Eskimo mtDNA Genome Reveals Matrilineal Discontinuity in Greenland." *Science* 320(5884):1787–1789.

Gilbert, M. T. P., L. Menez, R. C. Janaway, D. J. Tobin, A. Cooper, and A. S. Wilson. 2006. "Resistance of Degraded Hair Shafts to Contaminant DNA." *Forensic Science International* 156(2–3):208–212.

Gilbert, M. T. P., B. Shapiro, A. Drummond, and A. Cooper. 2005. "Post-mortem DNA Damage Hotspots in Bison (*Bison bison*) Provide Evidence for Both Damage and Mutational Hotspots in Human Mitochondrial DNA." *Journal of Archaeological Science* 32(7):1053–1060.

Gilbert, M. T. P., E. Willerslev, A. J. Hansen, I. Barnes, L. Rudbeck, N. Lynnerup, and A. Cooper. 2003. "Distribution Patterns of Postmortem Damage in Human Mitochondrial DNA." *American Journal of Human Genetics* 72(1):32–47.

Gill, P., P. L. Ivanov, C. Kimpton, R. Piercy, N. Benson, G. Tully, I. Evett, E. Hagelberg, and K. Sullivan. 1994. "Identification of the Remains of the Romanov Family by DNA Analysis." *Nature Genetics* 6(2):130–135.

Gitschier, J. 2008. Imagine: An Interview with Svante Pääbo. *PLoS Genetics* 4(3):e1000035.

Giuffra, E., J. M. H. Kijas, V. Amarger, Ö. Carlborg, J-T. Jeon, and L. Andersson. 2000. "The Origin of the Domestic Pig: Independent Domestication and Subsequent Introgression." *Genetics* 154(4):1785–1791.

Gobalet, K. W. 2001. "A Critique of Faunal Analysis: Inconsistency among Experts in Blind Tests." *Journal of Archaeological Science* 28(4):377–386.

Goebel, T., M. R. Waters, and D. H. O'Rourke. 2008. "The Late Pleistocene Dispersal of Modern Humans in the Americas." *Science* 319(5869):1497–1502.

Golson, J. 1977. "No Room at the Top: Agricultural Intensification in the New Guinea Highlands." In: J. Allen, J. Golson, and R. Jones, editors. *Sunda and Sahul: Prehistoric Studies in Southeast Asia, Melanesia and Australia.* London: Academic Press. pp. 601–638.

Gonder, M. K., H. M. Mortensen, F. A. Reed, A. de Sousa, and S. A. Tishkoff. 2007. "Whole-mtDNA Genome Sequence Analysis of Ancient African Lineages." *Molecular Biology and Evolution* 24(3):757–768.

Goodacre, S., A. Helgason, J. Nicholson, L. Southam, L. Ferguson, E. Hickey, E. Vega, K. Stefánsson, R. Ward, and B. Sykes. 2005. "Genetic Evidence for a Family-based Scandinavian Settlement of Shetland and Orkney during the Viking Periods." *Heredity* 95(2):129–135.

Götherström, A., C. Anderung, L. Hellborg, R. Elburg, E. C. Smith, D. G. Bradley, and H. Ellegren. 2005. "Cattle Domestication in the Near East was Followed by Hybridization with Aurochs Bulls in Europe." *Proceedings of the Royal Society B: Biological Sciences* 272(1579):2345–2351.

Götherström, A., M. J. Collins, A. Angerbjörn, and K. Lidén. 2002. "Bone Preservation and DNA Amplification." *Archaeometry* 404(3):395–404.

Götherström, A., C. Fischer, K. Linden. 1995. "X-raying Ancient Bone: A Destructive Method in Connection with DNA Analysis." *Laborativ Arkeologi* 8:26–28.

Graham-Rowe, D. 2012. "Sequence DNA in Seconds." *New Scientist* (25 Feb):23–24.

Gravlee, C. C. 2009. "How Race Becomes Biology: Embodiment of Social Inequality." *American Journal of Physical Anthropology* 139(1):47–57.

Gray, R. D. and F. M. Jordan. 2000. "Language Trees Support the Express-train Sequence of Austronesian Expansion." *Nature* 405:1052–1055.

Gray, R. D., A. J. Drummond, and S. J. Greenhill. 2009. Language Phylogenies Reveal Expansion Pulses and Pauses in Pacific Settlement." *Science* 323(5913):479–483.
Grayson, D. K. 1984. "Nineteenth-century Explanations of Pleistocene Extinctions: A Review and Analysis." In: P. S. Martin and R. D. Klein, editors. *Quaternary Extinctions: A Prehistoric Revolution*. Tucson: University of Arizona Press. pp. 5–39.
Grayson, D. K. 2001. "The Archaeological Record of Human Impacts on Animal Populations." *Journal of World Prehistory* 15(1):1–68.
Grayson, D. K. 2007. "Deciphering North American Pleistocene Extinctions." *Journal of Anthropological Research* 63(2):185–213.
Grayson, D. K. and D. J. Meltzer. 2002. "Clovis Hunting and Large Mammal Extinction: A Critical Review of the Evidence." *Journal of World Prehistory* 16(4):313–359.
Green, R. C. 1991. "Near and Remote Oceania—Disestablishing 'Melanesia' in Culture History." In: A. Pawley, editor. *Man and a Half: Essays in Pacific Anthropology and Ethnobiology in Honour of Ralph Bulmer*. Auckland: The Polynesian Society. pp. 491–502.
Green, R. E., A. W. Briggs, J. Krause, K. Prüfer, H. A. Burbano, M. Siebauer, M. Lachmann, and S. Pääbo. 2009. "The Neandertal Genome and Ancient DNA Authenticity." *The EMBO Journal* 28(17):2494–2502.
Green, R. E., J. Krause, A. W. Briggs, T. Maricic, U. Stenzel, M. Kircher, N. Patterson, H. Li, H. Zhai, M. H-Y. Fritz, and others. 2010. "A Draft Sequence of the Neandertal Genome." *Science* 328(5979):710–722.
Green, R. E., J. Krause, S. E. Ptak, A. W. Briggs, M. T. Ronan, J. F. Simons, L. Du, M. Egholm, J. M. Rothberg, M. Paunovic, and others. 2006. "Analysis of One Million Base Pairs of Neanderthal DNA." *Nature* 444(7117):330–336.
Green, R. E., A. S. Malaspinas, J. Krause, A. W. Briggs, P. L. F. Johnson, C. Uhler, M. Meyer, J. M. Good, T. Maricic, U. Stenzel, and others. 2008. "A Complete Neandertal Mitochondrial Genome Sequence Determined by High-throughput Sequencing." *Cell* 134(3):416–426.
Grieshaber, B. M., D. L. Osborne, A. F. Doubleday, and F. A. Kaestle. 2008. "A Pilot Study into the Effects of X-ray and Computed Exposure on the Amplification of DNA from Tomography Bone." *Journal of Archaeological Science* 35(3):681–687.
Grigson, C. 1991. "An African Origin for African Cattle?—Some Archaeological Evidence." *African Archaeological Review* 9(1):119–144.
Groves, C. 2001. "Lake Mungo 3 and his DNA." *Archaeology in Oceania* 36:166–167.
Gunnarsdóttir, E. D., M. Li, M. Bauchet, K. Finstermeier, and M. Stoneking. 2011. "High-Throughput Sequencing of Complete Human mtDNA Genomes from the Philippines." *Genome Research* 21:1–11.
Gutiérrez, G., D. Sánchez, and A. Marín. 2002. "A Reanalysis of the Ancient Mitochondrial DNA Sequences Recovered from Neandertal Bones." *Molecular Biology and Evolution* 19(8):1359–1366.
Haak, W., O. Balanovsky, J. J. Sanchez, S. Koshel, V. Zaporozhchenko, C. J. Adler, C. S. I. Der Sarkissian, G. Brandt, C. Schwarz, N. Nicklisch, and others. 2010. "Ancient DNA from European Early Neolithic Farmers Reveals Their Near Eastern Affinities." *PLoS Biology* 8(11):e1000536.
Haak, W., P. Forster, B. Bramanti, S. Matsumura, G. Brandt, M. Tänzer, R. Villems, C. Renfrew, D. Gronenborn, K. W. Alt, and others. 2005. "Ancient DNA from the First European Farmers in 7500-year-old Neolithic Sites." *Science* 310(5750):1016–1018.
Haas, C. J., A. Zink, E. Molńar, U. Szeimies, U. Reischl, A. Marcsik, Y. Ardagna, O. Dutour, G. Pálfi, and A. G. Nerlich. 2000. "Molecular Evidence for Different Stages of Tuberculosis in Ancient Bone Samples from Hungary." *American Journal of Physical Anthropology* 113(3):293–304.
Haas, C. J., A. Zink, G. Pálfi, U. Szeimies, and A. G. Nerlich. 2000. Detection of Leprosy in Ancient Human Skeletal Remains by Molecular Identification of *Mycobacterium leprae. American Journal of Clinical Pathology* 114(3):428–436.

Haber, M., D. E. Platt, D. A. Badro, Y. Xue, M. El-Sabai, M. A. Bonab, S. C. Youhanna, S. Saade, D. F. Soria-Hernandez, A. Royyuru, and others. 2011. "Influences of History, Geography, and Religion on Genetic Structure: The Maronites in Lebanon." *European Journal of Human Genetics* 19:334–340.

Hage, P. and J. Marck. 2003. "Matrilineality and the Melanesian Origin of Polynesian Y Chromosomes." *Current Anthropology* 44:S121–S127.

Hagelberg, E. and J. B. Clegg. 1993. "Genetic Polymorphisms in Prehistoric Pacific Islanders Determined by Analysis of Ancient Bone DNA." *Proceedings of the Royal Society B: Biological Sciences* 252(1334):163–170.

Hagelberg, E., S. Quevedo, D. Turbon, and J. B. Clegg. 1994. "DNA from Ancient Easter Islanders." *Nature* 369(6475):25–26.

Hagelberg, E., B. Sykes, and R. Hedges. 1989. "Ancient Bone DNA Amplified." *Nature* 342(6249):485.

Haile, J., D. G. Froese, R. D. Macphee, R. G. Roberts, L. J. Arnold, A. V. Reyes, M. Rasmussen, R. Nielsen, B. W. Brook, S. Robinson, and others. 2009. "Ancient DNA Reveals Late Survival of Mammoth and Horse in Interior Alaska." *Proceedings of the National Academy of Sciences of the United States of America* 106(52):22352–22357.

Hall, B. G. 2011. *Phylogenetic Trees Made Easy: A How to Manual.* Sunderland, CT: Sinauer Associates, Inc.

Hammer, M. F., T. Karafet, H. Park, K. Omoto, S. Harihara, M. Stoneking, and S. Horai. 2006. "Dual Origins of the Japanese: Common Ground for Hunter-gatherer and Farmer Y Chromosomes." *Journal of Human Genetics* 51(1):47–58.

Hammer, M. F. and S. L. Zegura. 2002. "The Human Y Chromosome Haplogroup Tree: Nomenclature and Phylogeography of its Major Divisions." *Annual Review of Anthropology* 31:303–321.

Handt, O., M. Höss, M. Krings, and S. Pääbo. 1994. "Ancient DNA: Methodological Challenges." *Cellular and Molecular Life Sciences* 50(6):524–529.

Handt, O., M. Richards, M. Trommsdorff, C. Kilger, J. Simanainen, O. Geogiev, K. Bauer, A. Stone, R. Hedges, W. Schaffner, and others. 1994. "Molecular Genetic Analyses of the Tyrolean Ice Man." *Science* 264(5166):1775–1778.

Hanotte, O., C. L. Tawah, D. G. Bradley, M. Okomo, Y. Verjee, J. Ochieng, and J. E. O. Rege. 2000. "Geographic Distribution and Frequency of a Taurine *Bos taurus* and an Indicine *Bos indicus* Y Specific Allele Amongst Sub-Saharan African Cattle Breeds." *Molecular Ecology* 9(4):387–396.

Hansen, A. J., E. Willerslev, C. Wiuf, T. Mourier, and P. Arctander. 2001. "Statistical Evidence for Miscoding Lesions in Ancient DNA Templates." *Molecular Biology and Evolution* 18(2):262–265.

Hardy, C., J. D. Vigne, D. Casañe, N. Dennebouy, J. C. Mounolou, and M. Monnerot. 1994. "Origin of European Rabbit (*Oryctolagus cuniculus*) in a Mediterranean Island—Zooarchaeology and Ancient DNA Examination." *Journal of Evolutionary Biology* 7(2):217–226.

Hartnup, K., L. Huynen, R. Te Kanawa, L. D. Shepherd, C. D. Millar, and D. M. Lambert. 2011. "Ancient DNA Recovers the Origins of Maori Feather Cloaks." *Molecular Biology and Evolution* 28(10):2741–2750.

Hasegawa, M., A. Di Rienzo, T. Kocher, and A. Wilson. 1993. "Toward a More Accurate Time Scale for the Human Mitochondrial DNA Tree." *Journal of Molecular Evolution* 37(4):347–354.

Hasegawa, M., H. Kishino, and T-A. Yano. 1985. "Dating of the Human-ape Splitting by a Molecular Clock of Mitochondrial DNA." *Journal of Molecular Evolution* 22(2):160–174.

Hasian, M. Jr. and E. Plec. 2002. "The Cultural, Legal, and Scientific Arguments in the Human Genome Diversity Debate." *Howard Journal of Communications* 13(4):301–319.

Hather, J. and P. V. Kirch. 1991. "Prehistoric Sweet-potato (*Ipomoea batatas*) from Mangaia Island, Central Polynesia." *Antiquity* 65(249):887–893.
Hatwell, J. N. and P. M. Sharp. 2000. "Evolution of Human Polyomavirus JC." *Journal of General Virology* 81:1191–1200.
Haynes, S., J. B. Searle, A. Bertman, and K. M. Dobney. 2002. "Bone Preservation and Ancient DNA: The Application of Screening Methods for Predicting DNA Survival." *Journal of Archaeological Science* 29(6):585–592.
Hebert, P. D. N., M. Y. Stoeckle, T. S. Zemlak, and C. M. Francis. 2004. "Identification of Birds through DNA Barcodes." *PLoS Biology* 2(10):1657–1663.
Helgason, A. and K. Stefánsson. 2003. "Erroneous Claims about the Impact of Mitochondrial DNA Sequence Database Errors." *American Journal of Human Genetics* 73(4):974–975.
Helgason, A., E. Hickey, S. Goodacre, V. Bosnes, K. Stefánsson, R. Ward, and B. Sykes. 2001. "MtDNA and the Islands of the North Atlantic: Estimating the Proportions of Norse and Gaelic Ancestry." *American Journal of Human Genetics* 68(3):723–737.
Helgason, A., C. Lalueza-Fox, S. Ghosh, S. Sigurðardóttir, M. L. Sampietro, E. Gigli, A. Baker, J. Bertranpetit, L. Árnadóttir, U. Þorsteinsdottir, and others. 2009. "Sequences from First Settlers Reveal Rapid Evolution in Icelandic mtDNA Pool." *PLoS Genetics* 5(1):e1000343.
Helgason, A., S. Sigurðardóttir, J. R. Gulcher, R. Ward, and K. Stefánsson. 2000. "MtDNA and the Origin of the Icelanders: Deciphering Signals of Recent Population History." *American Journal of Human Genetics* 66(3):999–1016.
Helmer, D. 1985. "Étude de la faune de Tell Assouad (Djezireh–Syrie) Sondage J. Cauvin." *Cahiers de l'Euphrate* 4:275–285.
Henn, B. M., L. R. Botigué, S. Gravel, W. Wang, A. Brisbin, J. K. Byrnes, K. Fadhlaoui-Zid, P. A. Zalloua, A. Moreno-Estrada, J. Bertranpetit, and others. 2012. "Genomic Ancestry of North Africans Supports Back-to-Africa Migrations." *PLoS Genetics* 8(1):e1002397.
Henn, B. M., C. R. Gignoux, M. Jobin, J. M. Granka, J. M. Macpherson, J. M. Kidd, L. Rodríguez-Botigué, S. Ramachandran, L. Hon, A. Brisbin, and others. 2011. "Hunter-gatherer Genomic Diversity Suggests a Southern African Origin for Modern Humans." *Proceedings of the National Academy of Sciences of the United States of America* 108(13):5154–5162.
Henn, B. M., C. R. Gignoux, A. A. Lin, P. J. Oefner, P. Shen, R. Scozzari, F. Cruciani, S. A. Tishkoff, J. L. Mountain, and P. A. Underhill. 2008. "Y-chromosomal Evidence of a Pastoral Migration through Tanzania to Southern Africa." *Proceedings of the National Academy of Sciences of the United States of America* 105(31):10693–10698.
Henry, D. O., P. F. Turnbull, A. Emery-Barbier, and A. Leroi-Gourhan. 1985. "Archaeological and Faunal Evidence from Natufian and Timnian Sites in Southern Jordan, with Notes on Pollen Evidence." *Bulletin of the American Schools of Oriental Research* 257:46–64.
Heyerdahl, T. 1952. *American Indians in the Pacific: The Theory behind the Kon-Tiki Expedition*. London: George Allen & Unwin, Ltd.
Hiendleder, S., B. Kaupe, R. Wassmuth, and A. Janke. 2002. "Molecular Analysis of Wild and Domestic Sheep Questions Current Nomenclature and Provides Evidence for Domestication from Two Different Subspecies." *Proceedings of the Royal Society B: Biological Sciences* 269(1494):893–904.
Hiendleder, S., K. Mainz, Y. Plante, and H. Lewalski. 1998. "Analysis of Mitochondrial DNA Indicates that Domestic Sheep are Derived from Two Different Ancestral Maternal Sources: No Evidence for Contributions from Urial and Argali Sheep." *Journal of Heredity* 89(2):113–120.
Hiendleder, S., S. H. Phua, and W. Hecht. 1999. "A Diagnostic Assay Discriminating between Two Major *Ovis aries* Mitochondrial DNA Haplogroups." *Animal Genetics* 30(3):211–213.

Higham, T., T. Compton, C. Stringer, R. Jacobi, B. Shapiro, E. Trinkhaus, B. Chandler, F. Gröning, C. Collins, S. Hillson, and others. 2011. "The Earliest Evidence for Anatomically Modern Humans in Northwestern Europe." *Nature* 479:521–524.

Higuchi, R., B. Bowman, M. Freiberger, O. A. Ryder, and A. C. Wilson. 1984. "DNA-Sequences from the Quagga, an Extinct Member of the Horse Family." *Nature* 312(5991):282–284.

Hill, A. V. S., D. K. Bowden, R. J. Trent, D. R. Higgs, S. J. Oppenheimer, S. L. Thein, K. N. P. Mickleson, D. J. Weatherall, and J. B. Clegg. 1985. "Melanesians and Polynesians Share a Unique Alpha-thalassemia Mutation." *American Journal of Human Genetics* 37(3):571–580.

Hill, A. V. S., D. Gentile, J. M. Bonnardot, J. Roux, D. J. Weatherall, and J. B. Clegg. 1987. "Polyneisan Origins and Affinities—Globin Gene Variants in Eastern Polynesia." *American Journal of Human Genetics* 40(5):453–463.

Hingston, M., S. M. Goodman, J. U. Ganzhorn, and S. Sommer. 2005. "Reconstruction of the Colonization of Southern Madagascar by Introduced *Rattus rattus*." *Journal of Biogeography* 32(9):1549–1559.

Hinkle, A. E. 2007. "Population Structure of Pacific *Cordyline fruticosa* (Laxmanniaceae) with Implications for Human Settlement of Polynesia." *American Journal of Botany* 94(5):828–839.

Ho, S. Y. W., T. H. Heupink, A. Rambaut, and B. Shapiro. 2007. "Bayesian Estimation of Sequence Damage in Ancient DNA." *Molecular Biology and Evolution* 24(6):1416–1422.

Hofreiter, M., C. Capelli, M. Krings, L. Waits, N. Conard, S. Münzel, G. Rabeder, D. Nagel, M. Paunovic, G. Jambresi, and others. 2002. "Ancient DNA Analyses Reveal High Mitochondrial DNA Sequence Diversity and Parallel Morphological Evolution of Late Pleistocene Cave Bears." *Molecular Biology and Evolution* 19(8):1244–1250.

Hofreiter, M., V. Jaenicke, D. Serre, A. von Haeseler, and S. Pääbo. 2001. "DNA Sequences from Multiple Amplifications Reveal Artifacts Induced by Cytosine Deamination in Ancient DNA." *Nucleic Acids Research* 29(23):4793–4799.

Hofreiter, M., O. Loreille, D. Ferriola, and T. J. Parsons. 2004. "Ongoing Controversy over Romanov Remains." *Science* 306(5695):407–408.

Hofreiter, M., S. Münzel, N. J. Conard, J. Pollack, M. Slatkin, G. Weiss, and S. Pääbo. 2007. "Sudden Replacement of Cave Bear Mitochondrial DNA in the Late Pleistocene." *Current Biology* 17(4):R122–R123.

Hofreiter, M., D. Serre, H. N. Poinar, M. Kuch, and S. Pääbo. 2001. "Ancient DNA." *Nature Reviews Genetics* 2(5):353–359.

Horai, S., K. Hayasaka, R. Kondo, K. Tsugane, and N. Takahata. 1995. "Recent African Origin of Modern Humans Revealed by Complete Sequences of Hominoid Mitochondrial DNA." *Proceedings of the National Academy of Sciences of the United States of America* 92(2):532–536.

Höss, M., P. Jaruga, T. Zastawny, M. Dizdaroglu, and S. Pääbo. 1996. "DNA Damage and DNA Sequence Retrieval from Ancient Tissues." *Nucleic Acids Research* 24:1304–1307.

Hoyer, B. H., N. W. van de Velde, M. Goodman, and R. B. Roberts. 1972. "Examination of Hominid Evolution by DNA Sequence Homology." *Journal of Human Evolution* 1(6):645–649.

Hublin, J-J. and R. G. Klein. 2011. "Northern Africa Could Also Have Housed the Source Population for Living Humans." *Proceedings of the National Academy of Sciences of the United States of America* 108(28):E227.

Hudjashov, G., T. Kivisild, P. A. Underhill, P. Endicott, J. J. Sanchez, A. A. Lin, P. Shen, P. Oefner, C. Renfrew, R. Villems, and others. 2007. "Revealing the Prehistoric Settlement of Australia by Y Chromosome and mtDNA Analysis." *Proceedings of the National Academy of Sciences of the United States of America* 104(21):8726–8730.

Huelsenbeck, J. P. and F. Ronquist. 2001. "MrBayes: Bayesian Inference of Phylogenetic Trees." *Bioinformatics* 17(8):754–755.

Huffman, T. N. 2007. *Handbook to the Iron Age: The Archaeology of Pre-Colonial Farming Societies in Southern Africa.* Scottsville, South Africa: University of KwaZulu-Natal Press.

Huffman, T. N. 2009. "Mapungubwe and Great Zimbabwe: The Origin and Spread of Social Complexity in Southern Africa." *Journal of Anthropological Archaeology* 28(1):37–54.

Human Genome Diversity Project. 1994. Summary Document (the Alghero Report). Available from http://hsblogs.stanford.edu/morrison/human-genome-diversity-project/

Human Genome Project. 2012. Human Genome Project Information. http://www.ornl.gov/sci/techresources/Human_Genome/home.shtml

Hunley, K., M. Dunn, E. Lindström, G. Reesink, A. Terrill, M. E. Healey, G. Koki, F. R. Friedlaender, and J. S. Friedlaender. 2008. "Genetic and Linguistic Coevolution in Northern Island Melanesia." *PLoS Genetics* 4(10):e1000239.

Hunley, K. L., M. E. Healy, and J. C. Long. 2009. "The Global Pattern of Gene Identity Variation Reveals a History of Long-range Migrations, Bottlenecks, and Local Mate Exchange: Implications for Biological Race." *American Journal of Physical Anthropology* 139(1):35–46.

Hurles, M. E., C. Irven, J. Nicholson, P. G. Taylor, F. R. Santos, J. Loughlin, M. A. Jobling, and B. C. Sykes. 1998. "European Y-chromosomal Lineages in Polynesians: A Contrast to the Population Structure Revealed by mtDNA." *American Journal of Human Genetics* 63(6):1793–1806.

Hurles, M. E., E. Maund, J. Nicholson, E. Bosch, C. Renfrew, B. C. Sykes, and M. A. Jobling. 2003. "Native American Y Chromosomes in Polynesia: The Genetic Impact of the Polynesian Slave Trade." *American Journal of Human Genetics* 72(5):1282–1287.

Huson, D. H. and D. Bryant. 2006. "Application of Phylogenetic Networks in Evolutionary Studies." *Molecular Biology and Evolution* 23(2):254–267.

Ijdo, J. W., A. Baldini, D. C. Ward, S. T. Reeders, R. A. Wells. 1991. "Origin of Human Chromosome 2: An Ancestral Telomere-telomere Fusion." *Proceedings of the National Academy of Sciences of the United States of America* 88(20):9051–9055.

Ingman, M. and U. Gyllensten. 2001. "Analysis of the Complete Human mtDNA Genome: Methodology and Inferences for Human Evolution." *Journal of Heredity* 92(6):454–461.

Ingman, M. and U. Gyllensten. 2003. "Mitochondrial Genome Variation and Evolutionary History of Australian and New Guinean Aborigines." *Genome Research* 13(7):1600–1606.

Ingman, M., H. Kaessmann, S. Pääbo, and U. Gyllensten. 2000. "Mitochondrial Genome Variation and the Origin of Modern Humans." *Nature* 408(6813):708–713.

International Human Genome Mapping Consortium. 2001. "A Physical Map of the Human Genome." *Nature* 409:934–941.

Ivanov, P. L., M. J. Wadhams, R. K. Roby, M. M. Holland, V. W. Weedn, and T. J. Parsons. 1996. "Mitochondrial DNA Sequence Heteroplasmy in the Grand Duke of Russia Georgij Romanov Establishes the Authenticity of the Remains of Tsar Nicholas II." *Nature Genetics* 12(4):417–420.

Ives, A. R., P. E. Midford, and T. Garland. 2007. "Within-species Variation and Measurement Error in Phylogenetic Comparative Methods." *Systematic Biology* 56(2):252–270.

Izawa, T., S. Konishi, A. Shomura, and M. Yano. 2009. "DNA Changes Tell Us About Rice Domestication." *Current Opinion in Plant Biology* 12(2):185–192.

Jaenicke-Després, V., E. S. Buckler, B. D. Smith, M. T. P. Gilbert, A. Cooper, J. Doebley, and S. Pääbo. 2003. "Early Allelic Selection in Maize as Revealed by Ancient DNA." *Science* 302(5648):1206–1208.

Jans, M. M. E., C. M. Nielsen-Marsh, C. I. Smith, M. J. Collins, and H. Kars. 2004. "Characterisation of Microbial Attack on Archaeological Bone." *Journal of Archaeological Science* 31(1):87–95.

Jansen, T., P. Forster, M. A. Levine, H. Oelke, M. Hurles, C. Renfrew, J. Weber, and K. Olek. 2002. "Mitochondrial DNA and the Origins of the Domestic Horse." *Proceedings of the National Academy of Sciences of the United States of America* 99(16):10905–10910.

Jin, H-J., C. Tyler-Smith, and W. Kim. 2009. "The Peopling of Korea Revealed by Analyses of Mitochondrial DNA and Y-chromosomal Markers." *PLoS One* 4(1):e4210.

Jobling, M. A. and C. Tyler-Smith. 1995. "Fathers and Sons: The Y Chromosome and Human Evolution." *Trends in Genetics* 11(11):449–456.

Jobling, M. A., M. E. Hurles, and C. Tyler-Smith. 2004. *Human Evolutionary Genetics: Origins, Peoples and Disease*. New York: Garland Science.

Jorgensen, T. H., H. N. Buttenschön, A. G. Wang, T. D. Als, A. D. Børglum, and H. Ewald. 2004. "The Origin of the Isolated Population of the Faroe Islands Investigated Using Y Chromosomal Markers." *Human Genetics* 115(1):19–28.

Jung, S., R. K. Duwal, and S. Lee. 2011. "COI Barcoding of True Bugs (Insecta, Heteroptera)." *Molecular Ecology Resources* 11(2):266–270.

Jungers, W. L., W. E. H. Harcourt-Smith, R. E. Wunderlich, M. W. Tocheri, S. G. Larson, T. Sutikna, R. A. Due, and M. J. Morwood. 2009. "The Foot of *Homo floresiensis*." *Nature* 459(7243):81–84.

Kacki, S., L. Rahalison, M. Rajerison, E. Ferroglio, and R. Bianucci. 2011. "Black Death in the Rural Cemetery of Saint-Laurent-de-la-Cabrerisse Aude-Languedoc, Southern France, 14th Century: Immunological Evidence." *Journal of Archaeological Science* 38(3):581–587.

Kadwell, M., M. Fernandez, H. F. Stanley, R. Baldi, J. C. Wheeler, R. Rosadio, and M. W. Bruford. 2001. "Genetic Analysis Reveals the Wild Ancestors of the Llama and the Alpaca." *Proceedings of the Royal Society B: Biological Sciences* 268(1485):2575–2584.

Kahn, P. 1994. "Genetic Diversity Project Tries Again." *Science* 266(5186):720–722.

Kaiser, C., B. Bachmeier, C. Conrad, A. Nerlich, H. Bratzke, W. Eisenmenger, and O. Peschel. 2008. "Molecular Study of Time Dependent Changes in DNA Stability in Soil Buried Skeletal Residues." *Forensic Science International* 177(1):32–36.

Karafet, T. M., F. L. Mendez, M. B. Meilerman, P. A. Underhill, S. L. Zegura, and H. F. Hammer. 2008. "New Binary Polymorphisms Reshape and Increase Resolution of the Human Y Chromosomal Haplogroup Tree." *Genome Research* 18(5):830–838.

Kasapidis, P., F. Suchentrunk, A. Magoulas, and G. Kotoulas. 2005. "The Shaping of Mitochondrial DNA Phylogeographic Patterns of the Brown Hare (*Lepus europaeus*) under the Combined Influence of Late Pleistocene Climatic Fluctuations and Anthropogenic Translocations." *Molecular Phylogenetics and Evolution* 34(1):55–66.

Kayser, M. 2010. "The Human Genetic History of Oceania: Near and Remote Views of Dispersal." *Current Biology* 20(4):R194–R201.

Kayser, M., S. Brauer, R. Cordaux, A. Casto, O. Lao, L. A. Zhivotovsky, C. Moyse-Faurie, R. B. Rutledge, W. Schiefenhövel, D. Gil, and others. 2006. "Melanesian and Asian Origins of Polynesians: mtDNA and Y Chromosome Gradients across the Pacific." *Molecular Biology and Evolution* 23(11):2234–2244.

Kayser, M., S. Brauer, G. Weiss, P. A. Underhill, L. Roewer, W. Schiefenhövel, and M. Stoneking. 2000. "Melanesian Origin of Polynesian Y Chromosomes." *Current Biology* 10(20):1237–1246.

Kayser, M., Y. Choi, M. van Oven, S. Mona, S. Brauer, R. J. Trent, D. Suarkia, W. Schiefenhövel, and M. Stoneking. 2008. "The Impact of the Austronesian Expansion: Evidence from mtDNA and Y Chromosome Diversity in the Admiralty Islands of Melanesia." *Molecular Biology and Evolution* 25(7):1362–1374.

Keller, A., A. Graefen, M. Ball, M. Matzas, V. Boisguerin, F. Maixner, P. Leidinger, C. Backes, R. Khairat, M. Forster, and others. 2012. "New Insights into the Tyrolean Iceman's Origin and Phenotype as Inferred by Whole-genome Sequencing." *Nature Communication* 3:698.

Kemp, B. M. and D. G. Smith. 2005. "Use of Bleach to Eliminate Contaminating DNA from the Surface of Bones and Teeth." *Forensic Science International* 154(1):53–61.

Khush, G. S. 1997. "Origin, Dispersal, Cultivation and Variation of Rice." *Plant Molecular Biology* 35(1–2):25–34.

Kijas, J. M. H. and L. Andersson. 2001. "A Phylogenetic Study of the Origin of the Domestic Pig Estimated from the Near-complete mtDNA Genome." *Journal of Molecular Evolution* 52(3):302–308.

Kim, U. K. and D. Drayna. 2005. "Genetics of Individual Differences in Bitter Taste Perception; Lessons from the PTC Gene." *Clinical Genetics* 67(4):275–280.

Kimura, B., F. B. Marshall, S. Y. Chen, S. Rosenbom, P. D. Moehlman, N. Tuross, R. C. Sabin, J. Peters, B. Barich, H. Yohannes, and others. 2010. "Ancient DNA from Nubian and Somali Wild Ass Provides Insights into Donkey Ancestry and Domestication." *Proceedings of the Royal Society B: Biological Sciences* 278(1702):50–57.

King, M. C. and A. C. Wilson. 1975. "Evolution at Two Levels in Humans and Chimpanzees." *Science* 188(4184):107–116.

Kirch, P. V. 2000. *On the Road of the Winds: An Archaeological History of the Pacific Islands before European Contact*. Berkeley: University of California Press.

Kirch, P. V. and R. C. Green. 2001. *Hawaiki, Ancestral Polynesia: An Essay in Historical Anthropology*. Cambridge: Cambridge University Press.

Kivisild, T., S. Rootsi, M. Metspalu, S. Mastana, K. Kaldma, J. Parik, E. Metspalu, M. Adojaan, H. V. Tolk, V. Stepanov, and others. 2003. "The Genetic Heritage of the Earliest Settlers Persists Both in Indian Tribal and Caste Populations." *American Journal of Human Genetics* 72(2):313–332.

Kivisild, T., H-V. Tolk, J. Parik, Y. Wang, S. Papiha, H-J. Bandelt, and R. Villems. 2002. "The Emerging Limbs and Twigs of the East Asian mtDNA Tree." *Molecular Biology and Evolution* 19(10):1737–1751.

Klein, R. G. 2009. *The Human Career: Human Biological and Cultural Origins*. Chicago: The University of Chicago Press.

Knapp, M. and M. Hofreiter. 2010. "Next Generation Sequencing of Ancient DNA: Requirements, Strategies and Perspectives." *Genes* 1(2):227–243.

Knapp, M., A. C. Clarke, K. A. Horsburgh, E. A. Matisoo-Smith. 2012. "Setting the Stage—Building and Working in an Ancient DNA Laboratory." *Annals of Anatomy* 194(2012):3–6.

Knapp, M., N. Rohland, J. Weinstock, G. Baryshnikov, A. Sher, D. Nagel, G. Rabeder, R. Pinhasi, H. A. Schmidt, and M. Hofreiter. 2009. "First DNA Sequences from Asian Cave Bear Fossils Reveal Deep Divergences and Complex Phylogeographic Patterns." *Molecular Ecology* 18(6):1225–1238.

Knight, A., L. A. Zhivotovsky, D. H. Kass, D. E. Litwin, L. D. Green, P. S. White, and J. L. Mountain. 2004. "Molecular, Forensic and Haplotypic Inconsistencies Regarding the Identity of the Ekaterinburg Remains." *Annals of Human Biology* 31(2):129–138.

Koch, P. L. and A. D. Barnosky. 2006. "Late Quaternary Extinctions: State of the Debate." *Annual Review of Ecology, Evolution, and Systematics* 37:215–250.

Kocher, T. D., W. K. Thomas, A. Meyer, S. V. Edwards, S. Pääbo, F. X. Villablanca, and A. C. Wilson. 1989. "Dynamics of the Mitochondrial-DNA Evolution in Animals—Amplification and Sequencing with Conserved Primers." *Proceedings of the National Academy of Sciences of the United States of America* 86(16):6196–6200.

Kochzius, M., C. Seidel, A. Antoniou, S. K. Botla, D. Campo, A. Cariani, E. G. Vazquez, J. Hauschild, C. Hervet, S. Hjörleifsdottir, and others. 2010. "Identifying Fishes through DNA Barcodes and Microarrays." *PLoS One* 5(9):e12620.

Kolman, C. J., A. Centurion-Lara, S. A. Lukehart, D. W. Owsley, and N. Tuross. 1999. "Identification of *Treponema pallidum* Subspecies Pallidum in a 200-year-old Skeletal Specimen." *Journal of Infectious Diseases* 180:2060–2063.

Koon, H. E.C., T. P. O'Connor, and M. J. Collins. 2010. "Sorting the Butchered from the Boiled." *Journal of Archaeological Science* 37(1):62–69.

Kovach, M. J., M. T. Sweeney, and S. R. McCouch. 2007. "New Insights into the History of Rice Domestication." *Trends in Genetics* 23(11):578–587.

Krause, J., A. W. Briggs, M. Kircher, T. Maricic, N. Zwyns, A. Derevianko, and S. Pääbo. 2010. "A Complete mtDNA Genome of an Early Modern Human from Kostenki, Russia." *Current Biology* 20(3):231–236.

Krause, J., Q. M. Fu, J. M. Good, B. Viola, M. V. Shunkov, A. P. Derevianko, and S. Pääbo. 2010. "The Complete Mitochondrial DNA Genome of an Unknown Hominin from Southern Siberia." *Nature* 464(7290):894–897.

Krings, M., H. Geisert, R. W. Schmitz, H. Krainitzki, and S. Pääbo. 1999. "DNA Sequence of the Mitochondrial Hypervariable Region II from the Neandertal Type Specimen." *Proceedings of the National Academy of Sciences of the United States of America* 96(10):5581–5585.

Krings, M., A. Stone, R. W. Schmitz, H. Krainitszki, M. Stoneking, and S. Pääbo. 1997. Neandertal DNA Sequences and the Origin of Modern Humans. *Cell* 90(1):19–30.

Kumar, S., M. Nagarajan, J. S. Sandhu, N. Kumar, and V. Behl. 2007. "Phylogeography and Domestication of Indian River Buffalo." *BMC Evolutionary Biology* 7:186.

Kumar, S., M. Nei, J. Dudley, and K. Tamura. 2008. "MEGA: A Biologist-centric Software for Evolutionary Analysis of DNA and Protein Sequences." *Briefings in Bioinformatics* 9(4):299–306.

Kumar, S., P. B. S. V. Padmanabham, R. R. Ravuri, K. Uttaravalli, P. Koneru, P. A. Mukherjee, B. Das, M. Kotal, D. Xaviour, S. Y. Saheb, and others. 2008. "The Earliest Settlers' Antiquity and Evolutionary History of Indian Populations: Evidence from M2 mtDNA Lineage." *BMC Evolutionary Biology* 8(1):230.

Kumar, V., A. N. S. Reddy, J. P. Babu, T. N. Rao, B. T. Langstieh, K. Thangaraj, A. G. Reddy, L. Singh, and B. M. Reddy. 2007. "Y-chromosome Evidence Suggests a Common Paternal Heritage of Austro-Asiatic Populations." *BMC Evolutionary Biology* 7(1):47.

Lahr, M. M. and R. A. Foley. 1998. "Towards a Theory of Modern Human Origins: Geography, Demography, and Diversity in Recent Human Evolution." *Yearbook of Physical Anthropology* 41:137–176.

Lalueza-Fox, C., E. Gigli, M. de la Rasilla, J. Fortea, and A. Rosas. 2009. "Bitter Taste Perception in Neanderthals through the Analysis of the TAS2R38 Gene." *Biology Letters* 5(6):809–811.

Lalueza-Fox, C., E. Gigli, M. de la Rasilla, J. Fortea, A. Rosas, J. Bertranpetit, and J. Krause. 2008. "Genetic Characterization of the ABO Blood Group in Neandertals." *BMC Evolutionary Biology* 8:342.

Lalueza-Fox, C., H. Römpler, D. Caramelli, C. Stäubert, G. Catalano, D. Hughes, N. Rohland, E. Pilli, L. Longo, S. Condemi, and others. 2007. "A Melanocortin 1 Receptor Allele Suggests Varying Pigmentation among Neanderthals." *Science* 318(5855):1453–1455.

Lansman, R. A. and D. A. Clayton. 1975. "Selective Nicking of Mammalian Mitochondrial-DNA In vivo—Photosensitization by Incorporation of 5-Bromodeoxyuridine." *Journal of Molecular Biology* 99(4):761–776.

Lari, M., E. Rizzi, L. Milani, G. Corti, C. Balsamo, S. Vai, G. Catalano, E. Pilli, L. Longo, S. Condemi, and others. 2010. "The Microcephalin Ancestral Allele in a Neanderthal Individual." *PLoS One* 5(5):e10648.

Larson, G., U. Albarella, K. Dobney, P. Rowley-Conwy, J. Schibler, A. Tresset, J. D. Vigne, C. J. Edwards, A. Schlumbaum, A. Dinu, and others. 2007. "Ancient DNA, Pig Domestication, and the Spread of the Neolithic into Europe." *Proceedings of the National Academy of Sciences of the United States of America* 104(39):15276–15281.

Larson, G., T. Cucchi, M. Fujita, E. Matisoo-Smith, J. Robins, A. Anderson, B. Rolett, M. Spriggs, G. Dolman, T-H. Kim, and others. 2007. "Phylogeny and Ancient DNA of *Sus* Provides Insights into Neolithic Expansion in Island Southeast Asia and Oceania." *Proceedings of the National Academy of Sciences of the United States of America* 104(12):4834–4839.

Larson, G., K. Dobney, U. Albarella, M. Fang, E. Matisoo-Smith, J. Robins, S. Lowden, H. Finlayson, T. Brand, E. Willerslev, and others. 2005. "Worldwide Phylogeography of Wild Boar Reveals Multiple Centers of Pig Domestication." *Science* 307(5715):1618–1621.

Larson, S. G., W. L. Jungers, M. J. Morwood, T. Sutikna, E. Jatmiko, W. Saptomo, R. A. Due, and T. Djubiantono. 2007. "*Homo floresiensis* and the Evolution of the Hominin Shoulder." *Journal of Human Evolution* 53(6):718–731.

Lee, T., J. B. Burch, T. Coote, B. Fontaine, O. Gargominy, P. Pearce-Kelly, and D. Ó. Foighil. 2007. "Prehistoric Inter-archipelago Trading of Polynesian Tree Snails Leaves a Conservation Legacy." *Proceedings of the Royal Society B: Biological Sciences* 274(1627):2907–2914.

Legge, A. J. 1975. "The Fauna of Tell Abu Hureyra: Preliminary Analysis." *Proceedings of the Prehistoric Society* 41:74–77.

Legge, A. J. 1977. "Origins of Agriculture in the Near East." In: V. Megaw, editor. *Hunters, Gatherers and First Farmers beyond Europe*. Leceister: University Press. pp. 51–76.

Legge, T. 1996. "The Beginning of Caprine Domestication in Southwest Asia." In: D. R. Harris, editor. *The Origins and Spread of Agriculture and Pastoralism in Eurasia*. London: University College London Press. pp. 238–262.

Leonard, J. A., R. K. Wayne, J. Wheeler, R. Valadez, S. Guillén, and C. Vilà. 2002. "Ancient DNA Evidence for Old World Origin of New World Dogs." *Science* 298(5598):1613–1616.

Li, H., X. Cai, E. R. Winograd-Cort, B. Wen, X. Cheng, Z. Qin, W. Liu, Y. Liu, S. Pan, J. Qian, and others. 2007. "Mitochondrial DNA Diversity and Population Differentiation in Southern East Asia." *American Journal of Physical Anthropology* 134(4):481–488.

Linz, B. and S. C. Schuster. 2007. "Genomic Diversity in *Helicobacter* and Related Organisms." *Research in Microbiology* 158(10):737–744.

Liti, G., D. M. Carter, A. M. Moses, J. Warringer, L. Parts, S. A. James, R. P. Davey, I. N. Roberts, A. Burt, V. Koufopanou, and others. 2009. "Population Genomics of Domestic and Wild Yeasts." *Nature* 458:337–341.

Liu, H., F. Prugnolle, A. Manica, and F. Balloux. 2006. "A Geographically Explicit Genetic Model of Worldwide Human-settlement History." *American Journal of Human Genetics* 79(2):230–237.

Liu, Y-P., G-S. Wu, Y-G. Yao, Y-W. Miao, G. Luikart, M. Baig, A. Beja-Pereira, Z-L. Ding, M. G. Palanichamy, and Y. P. Zhang. 2006. "Multiple Maternal Origins of Chickens: Out of the Asian Jungles." *Molecular Phylogenetics and Evolution* 38(1):12–19.

Loftus, R. T., D. E. MacHugh, D. G. Bradley, P. M. Sharp, and P. Cunningham. 1994. "Evidence for Two Independent Domestications of Cattle." *Proceedings of the National Academy of Sciences of the United States of America* 91(7):2757–2761.

Lone Dog, L. 1999. "Whose Genes are They? The Human Genome Diversity Project." *Journal of Health and Social Policy* 10(4):51–66.

Lopes, M. S., D. Mendonça, M. R. dos Santos, J. E. Eiras-Dias, and A. da Câmara Machado. 2009. "New Insights on the Genetic Basis of Portuguese Grapevine and on Grapevine Domestication." *Genome* 52(9):790–800.

Loreille, O., J. D. Vigne, C. Hardy, C. Callou, F. Treinen Claustre, N. Dennebouy, and M. Monnerot. 1997. "First Distinction of Sheep and Goat Archaeological Bones by Means of Their Fossil mtDNA." *Journal of Archaeological Science* 24(1):33–37.

Ludwig, A., M. Pruvost, M. Reissmann, N. Benecke, G. A. Brockmann, P. Castaños, M. Cieslak, S. Lippold, L. Llorente, A. S. Malaspinas, and others. 2009. "Coat Color Variation at the Beginning of Horse Domestication." *Science* 324(5926):485.

Luikart, G., L. Gielly, L. Excoffier, J-D. Vigne, J. Bouvet, and P. Taberlet. 2001. "Multiple Maternal Origins and Weak Phylogeographic Structure in Domestic Goats." *Proceedings of the National Academy of Sciences of the United States of America* 98(10):5927–5932.

Lum, J. K. and R. L. Cann. 2000. "MtDNA Lineage Analyses: Origins and Migrations of Micronesians and Polynesians." *American Journal of Physical Anthropology* 113(2):151–168.

Lum, J. K., A. Kaneko, K. Tanabe, N. Takahashi, A. Björkman, and T. Kobayakawa. 2004. "Malaria Dispersal among Islands: Human Mediated *Plasmodium falciparum* Gene Flow in Vanuatu, Melanesia." *Acta Tropica* 90(2):181–185.

Lyapunova, E. A., T. B. Bunch, N. N. Voronsov, and R. S. Hoffmann. 1997. "Chromosome Sets and the Taxonomy of Severtsov's Wild Sheep (*Ovis ammon severtzovi*)." *Russian Journal of Zoology* 1:387–396.

Lydekker, R. 1912. *The Sheep and its Cousins*. London: Allen.

Maca-Meyer, N., A. M. González, J. M. Larruga, C. Flores, and V. M. Cabrera. 2001. "Major Genomic Mitochondrial Lineages Delineate Early Human Expansions." *BMC Genetics* 2:13.

Macaulay, V., C. Hill, A. Achilli, C. Rengo, D. Clarke, W. Meehan, J. Blackburn, O. Semino, R. Scozzari, F. Cruciani, and others. 2005. "Single, Rapid Coastal Settlement of Asia Revealed by Analysis of Complete Mitochondrial Genomes." *Science* 308(5724):1034–1036.

MacHugh, D. E., M. D. Shriver, R. T. Loftus, P. Cunningham, and D. G. Bradley. 1997. "Microsatellite DNA Variation and the Evolution, Domestication and Phylogeography of Taurine and Zebu Cattle (*Bos taurus* and *Bos indicus*)." *Genetics* 146(3):1071–1086.

Maddison, D. R. 1991. "African Origin of Human Mitochondrial DNA Reexamined." *Systematic Zoology* 40(3):355–363.

Maddison, D. R., M. Ruvolo, and D. L. Swofford. 1992. "Geographic Origins of Human Mitochondrial DNA: Phylogenetic Evidence from Control Region Sequences." *Systematic Biology* 41(1):111–124.

Maderspacher, F. 2008. "Ötzi." *Current Biology* 18(21):R990–R991.

Maji, S., S. Krithika, and T. S. Vasulu. 2009. "Phylogeographic Distribution of Mitochondrial DNA Macrohaplogroup M in India." *Journal of Genetics* 88(1):127–139.

Majumder, P. P. 2010. "The Human Genetic History of South Asia." *Current Biology* 20(4):R184–R187.

Malhi, R. S., B. M. Kemp, J. A. Eshleman, J. Cybulski, D. G. Smith, S. Cousins, and H. Harry. 2007. "Mitochondrial Haplogroup M Discovered in Prehistoric North Americans." *Journal of Archaeological Science* 34(4):642–648.

Malmström, H., E. M. Svensson, M. T. P. Gilbert, E. Willerslev, A. Götherström, and G. Holmlund. 2007. "More on Contamination: The Use of Asymmetric Molecular Behavior to Identify Authentic Ancient Human DNA." *Molecular Biology and Evolution* 24(4):998–1004.

Malmström, H., C. Vilà, M. T. P. Gilbert, J. Stora, E. Willerslev, G. Holmlund, A. Götherström. 2008. "Barking Up the Wrong Tree: Modern Northern European Dogs Fail to Explain Their Origin." *BMC Evolutionary Biology* 8(71).

Malyarchuk, B., T. Grzybowski, M. Derenko, M. Perkova, T. Vanecek, J. Lazur, P. Gomolcak, and I. Tsybovsky. 2008. "Mitochondrial DNA Phylogeny in Eastern and Western Slavs." *Molecular Biology and Evolution* 25(8):1651–1658.

Mannucci, A., K. M. Sullican, P. L. Ivanov, and P. Gill. 1994. "Forensic Application of a Rapid and Quantitative DNA Sex Test by Amplification of the X-Y Homologous Gene Amelogenin." *International Journal of Legal Medicine* 106:190–193.

Marean, C. W., M. Bar-Matthews, J. Bernatchez, E. Fisher, P. Goldberg, A. I. R. Herries, Z. Jacobs, A. Jerardino, P. Karkanas, T. Minichillo, and others. 2007. Early "Human Use of Marine Resources and Pigment in South Africa during the Middle Pleistocene." *Nature* 449:905–908.

Maricic, T., M. Whitten, and S. Pääbo. 2010. "Multiplexed DNA Sequence Capture of Mitochondrial Genomes Using PCR Products." *Plos One* 5(11).

Marks, J. 1995. "Anthropology and Race." *Nature* 377(6550):570.

Martin, P. S. 1984. "Prehistoric Overkill: The Global Model." In: P. S. Martin and R. G. Klein, editors. *Quaternary Extinctions: A Prehistoric Revolution*. Tucson: University of Arizona Press. pp. 354–403.

Martinez, A. M. and O. C. Hamsici. 2008. "Who is LB1? Discriminant Analysis for the Classification of Specimens." *Pattern Recognition* 41(11):3436–3441.

Martínková, N., R. A. McDonald, and J. B. Searle. 2007. "Stoats (*Mustela erminea*) Provide Evidence of Natural Overland Colonization of Ireland." *Proceedings of the Royal Society B: Biological Sciences* 274(1616):1387–1393.

Mason, I. L. 1996. *A World Dictionary of Livestock Breeds, Types and Varieties*. Oxford: CAB International.

Matheson, C. D., K. K. Vernon, A. Lahti, R. Fratpietro, M. Spigelman, S. Gibson, C. L. Greenblatt, and H. D. Donoghue. 2009. "Molecular Exploration of the First-century Tomb of the Shroud in Akeldama, Jerusalem." *PLoS One* 4(12): e8319.

Matisoo-Smith, E. and J. Robins. 2009. "Mitochondrial DNA Evidence for the Spread of Pacific Rats through Oceania." *Biological Invasions* 11(7):1521–1527.

Matisoo-Smith, E. and J. H. Robins. 2004. "Origins and Dispersals of Pacific Peoples: Evidence from mtDNA Phylogenies of the Pacific Rat." *Proceedings of the National Academy of Sciences of the United States of America* 101(24):9167–9172.

Matisoo-Smith, E., M. Hingston, G. Summerhayes, J. Robins, H. A. Ross, and M. Hendy. 2009. "On the Rat Trail in Near Oceania: Applying the Commensal Model to the Question of the Lapita Colonization." *Pacific Science* 63(4):465–475.

Matthews, P. J. 1990. *The Origins, Dispersal and Domestication of Taro*. Canberra: The Australian National University.

May, K. J. and J. B. Ristaino. 2004. "Identity of the mtDNA Haplotype(s) of *Phytophthora infestans* in Historical Specimens from the Irish Potato Famine." *Mycological Research* 108:471–479.

McCain, L. 2002. "Informing Technology Policy Decisions: The US Human Genome Project's Ethical, Legal, and Social Implications Programs as a Critical Case." *Technology in Society* 24(1–2):111–132.

McCracken, R. D. 1971. "Lactase Deficiency: An Example of Dietary Evolution." *Current Anthropology* 12(4/5):479–517.

McPherson, J. D., M. Marra, L. Hillier, R. H. Waterston, A. Chinwalla, J. Wallis, M. Sekhon, K. Wylie, E. R. Mardis, R. K. Wilson, and others. 2001. "A Physical Map of the Human Genome." *Nature* 409(6822):934–941.

Meadow, R. H. 1984. "Animal Domestication in the Middle East: A View from the Eastern Margin." In: J. Clutton-Brock and C. Grigson, editors. Animals and Archaeology: 3. *Early Herders and Their Flocks*. Oxford: Oxford University Press. pp. 309–337.

Meadow, R. H. 1993. "Animal Domestication in the Middle East: A Revised View from the Eastern Margin." In: G. Possehl, editor. *Harappan Civilisation*. New Delhi: Oxford and IBH. pp. 295–320.

Meadows, J. R. S., I. Cemal, O. Karaca, E. Gootwine, and J. W. Kijas. 2007. "Five Ovine Mitochondrial Lineages Identified from Sheep Breeds of the Near East." *Genetics* 175(3):1371–1379.

Meadows, J. R. S., S. Hiendleder, and J. W. Kijas. 2011. "Haplogroup Relationships between Domestic and Wild Sheep Using a Mitogenome Panel." *Heredity* 106:700–706.

Melchior, L., M. T. P. Gilbert, T. Kivisild, N. Lynnerup, and J. Dissingl. 2008. "Rare mtDNA Haplogroups and Genetic Differences in Rich and Poor Danish Iron-age Villages." *American Journal of Physical Anthropology* 135(2):206–215.

Melé, M., A. Javed, M. Pybus, P. Zalloua, M. Haber, D. Comas, M. G. Netea, O. Balanovsky, E. Balanovska, L. Jin, and others. 2012. "Recombination Gives a New

Insight in the Effective Populations Size and the History of the Old World Human Populations." *Molecular Biology and Evolution* 29(1):25–30.

Mellars, P. 2006. "Going East: New Genetic and Archaeological Perspectives on the Modern Human Colonisation of Eurasia." *Science* 313(5788):796–800.

Melton, T., R. Peterson, A. J. Redd, N. Saha, A. S. M. Sofro, J. Martinson, and M. Stoneking. 1995. "Polynesian Genetic Affinities with Southeast-Asian Populations as Identified by mtDNA Analysis." *American Journal of Human Genetics* 57(2):403–414.

Mendel, G. 1865. "Experiments in Plant Hybridization." *Verhandlungen des Naturforschenden Vereines in Brünn* Bd IV:3–47.

Metspalu, M., T. Kivisild, E. Metspalu, J. Parik, G. Hudjashov, K. Kaldma, P. Serk, M. Karmin, D. Behar, M. T. Gilbert, and others. 2004. "Most of the Extant mtDNA Boundaries in South and Southwest Asia Were Likely Shaped during the Initial Settlement of Eurasia by Anatomically Modern Humans." *MC Genetics* 5(1):26.

Meudt, H. M. and A. C. Clarke. 2007. "Almost Forgotten or Latest Practice? AFLP Applications, Analyses and Advances." *Trends in Plant Science* 12(3):106–117.

Meyer, A. and A. C. Wilson. 1990. "Origin of Tetrapods Inferred from their Mitochondrial DNA Affiliation to Lungfish." *Journal of Molecular Evolution* 31(5):359–364.

Meyer, A., T. D. Kocher, and A. C. Wilson. 1988. "Mitochondrial Genes Amplifed via the Polymerase-Chain-Reaction—Inferences for the Evolution of Cichlid Fishes from Mitochondrial Gene Sequences." *American Zoologist* 28(4):A35.

Meyer, A., T. D. Kocher, and A. C. Wilson. 1989. "Speciation Rates and Evolutionary Relationships of East-African Cichlid Fishes Inferred from mtDNA Sequences." *American Zoologist* 29(4):A32.

Meyer, A., T. D. Kocher, P. Basasibwaki, and A. C. Wilson. 1990. "Monophyletic Origin of Lake Victoria Cichlid Fishes Suggested by Mitochondrial-DNA Sequences." *Nature* 347(6293):550–553.

Mikkelsen, T. S., L. W. Hillier, E. E. Eichler, M. C. Zody, D. B. Jaffe, S. P. Yang, E. Enard, I. Hellmann, K. Lindblad-Toh, T. K. Altheide, and others. 2005. "Initial Sequence of the Chimpanzee Genome and Comparison with the Human Genome." *Nature* 437(7055):69–87.

Millar, C. D., L. Huynen, S. Subramanian, E. Mohandesan, and D. M. Lambert. 2008. "New Developments in Ancient Genomics." *Trends in Ecology and Evolution* 23(7):386–393.

Miller, G. H., M. L. Fogel, J. W. Magee, M. K. Gagan, S. J. Clarke, and B. J. Johnson. 2005. "Ecosystem Collapse in Pleistocene Australia and a Human Role in Megafaunal Extinction." *Science* 309(5732):287–290.

Miranda, J. J., T. Takasaka, H. Y. Zheng, T. Kitamura, and Y. Yogo. 2004. "JC Virus Genotype Profile in the Mamanwa, a Philippine Negrito Tribe, and Implications for its Population History." *Anthropological Science* 112(2):173–178.

Mitchell, P. 2002. *The Archaeology of Southern Africa*. Cambridge: Cambridge University Press.

Mitchell, P. 2010. "Genetics and Southern African Prehistory: An Archaeological View." *Journal of Anthropological Sciences* 88:73–92.

Modiano, G., B. M. Ciminelli, and P. F. Pignatti. 2007. "Cystic Fibrosis and Lactase Persistence: A Possible Correlation." *European Journal of Human Genetics* 15(3):255–259.

Mona, S., K. E. Grunz, S. Brauer, B. Pakendorf, L. Castrì, H. Sudoyo, S. Marzuki, R. H. Barnes, J. Schmidtke, M. Stoneking, and others. 2009. "Genetic Admixture History of Eastern Indonesia as Revealed by Y-chromosome and Mitochondrial DNA Analysis." *Molecular Biology and Evolution* 26(8):1865–1877.

Montano, V., G. Ferri, V. Marcari, C. Batini, O. Anyaele, G. Destro-Bisol, and D. Comas. 2011. "The Bantu Expansion Revisited: A New Analysis of Y Chromosome Variation in Central Western Africa." *Molecular Ecology* 20(13):2693–2708.

Mooder, K. P., T. G. Schurr, F. J. Bamforth, V. I. Bazahiski, and N. A. Savel'ev. 2006. "Population Affinities of Neolithic Siberians: A Snapshot from Prehistoric Lake Baikal." *American Journal of Physical Anthropology* 129(3):349–361.

Moodley, Y., B. Linz, Y. Yamaoka, H. M. Windsor, S. Breurec, J. Y. Wu, A. Maady, S. Bernhoft, J. M. Thiberge, S. Phuanukoonnon, and others. 2009. "The Peopling of the Pacific from a Bacterial Perspective." *Science* 323(5913):527–530.

Moorjani, P., N. Patterson, J. N. Hirschhorn, A. Keinan, L. Hao, G. Atzmon, E. Burns, H. Ostrer, A. L. Price, and D. Reich. 2011. "The History of African Gene Flow into Southern Europeans, Levantines, and Jews." *PLoS Genetics* 7(4): e1001373.

Morwood, M. J. and W. L. Jungers. 2009. "Conclusions: Implications of the Liang Bua Excavations for Hominin Evolution and Biogeography." *Journal of Human Evolution* 57(5):640–648.

Morwood, M. J., P. B. O'Sullivan, F. Aziz, and A. Raza. 1998. "Fission-track Ages of Stone Tools and Fossils on the East Indonesian Island of Flores." *Nature* 392(6672):173–176.

Morwood, M. J., R. P. Soejono, R. G. Roberts, T. Sutikna, C. S. M. Turney, K. E. Westaway, W. J. Rink, J. X. Zhao, G. D. van den Bergh, R. A. Due, and others. 2004. "Archaeology and Age of a New Hominin from Flores in Eastern Indonesia." *Nature* 431(7012):1087–1091.

Morwood, M. J., T. Sutikna, E. W. Saptomo, Jatmiko, D. R. Hobbs, and K. E. Westaway. 2009. "Preface: Research at Liang Bua, Flores, Indonesia." *Journal of Human Evolution* 57(5):437–449.

Mullis, K., F. Faloona, S. Scharf, R. Saiki, G. Horn, and H. Erlich. 1986. "Specific Enzymatic Amplification of DNA In Vitro: The Polymerase Chain Reaction." *Cold Spring Harbor Symposia on Quantitative Biology* 51:263–273.

Mullis, K. B. 1990. "The Unusual Origin of the Polymerase Chain-reaction." *Scientific American* 262(4):56–65.

Mullis, K. B. and F. A. Faloona. 1987. "Specific Synthesis of DNA Invitro via a Polymerase-Catalyzed Chain-Reaction." *Methods in Enzymology* 155:335–350.

Naderi, S., H-R. Resaei, F. Pompanon, M. C. B. Blum, R. Negrini, H-R. Naghash, Ö. Balkız, M. Mashkour, O. E. Gaggiotti, P. Ajmone-Marsan, and others. 2008. "The Goat Domestication Process Inferred from Large-scale Mitochondrial DNA Analysis of Wild and Domestic Individuals." *Proceedings of the National Academy of Sciences of the United States of America* 105(46):17659–17664.

National Research Council (NRC) (US). 1997. *Evaluating Human Diversity*. Washington DC: National Academy Press.

Nei, M. and S. Kumar. 2000. *Molecular Evolution and Phylogenetics*. New York: Oxford University Press.

Nicholls, A., E. Matisoo-Smith, and M. S. Allen. 2003. "A Novel Application of Molecular Techniques to Pacific Archaeofish Remains." *Archaeometry* 45(1):133–147.

Nicholls, H. 2005. "Ancient DNA Comes of Age." *PLoS Biology* 3(2):192–196.

Nicholson, G. J., J. Tomiuk, A. Czarnetzki, L. Bachmann, and C. M. Pusch. 2002. "Detection of Bone Glue Treatment as a Major Source of Contamination in Ancient DNA Analyses." *American Journal of Physical Anthropology* 118:117–120.

O'Connell, J. F. and J. Allen. 1998. "When did Humans First Arrive in Greater Australia and Why is it Important to Know?" *Evolutionary Anthropology* 6(4):132–146.

O'Connell, J. F. and J. Allen. 2004. "Dating the Colonization of Sahul (Pleistocene Australia-New Guinea): A Review of Recent Research." *Journal of Archaeological Science* 31(6):835–853.

O'Rourke, D. H. and J. A. Raff. 2010. "The Human Genetic History of the Americas: The Final Frontier." *Current Biology* 20(4):R202–R207.

Outram, A. K., N. A. Stear, R. Bendrey, S. Olsen, A. Kasparov, V. Zaibert, N. Thorpe, and R. P. Evershed. 2009. "The Earliest Horse Harnessing and Milking." *Science* 323(5919):1332–1335.

Ovchinnikov, I. V., A. Götherström, G. P. Romanova, V. M. Kharitonov, K. Lidén, and W. Goodwin. 2000. "Molecular Analysis of Neanderthal DNA from the Northern Caucasus." *Nature* 404(6777):490–493.

Owings, A. C., M. C. Dulik, S. I. Zhadanov, J. B. Gaieski, J. Ramos, M. B. Moss, F. Natkong, T. G. Schurr, and The Genographic C. 2011. "Investigating Athapaskan History through the Analysis of Genetic Variation in the Tlingit and Haida Populations of Southeast Alaska." *American Journal of Physical Anthropology* 144:232.

Pääbo, S. 1985. "Molecular Cloning of Ancient Egyptian Mummy DNA." *Nature* 314(6012):644–645.

Pääbo, S. and A. C. Wilson. 1991. "Miocene DNA Sequences—A Dream Come True?" *Current Biology* 1(1):45–46.

Pääbo, S., H. Poinar, D. Serre, V. Jaenicke-Després, J. Hebler, N. Rohland, M. Kuch, J. Krause, L. Vigilant, and M. Hofreiter. 2004. "Genetic Analyses from Ancient DNA." *Annual Review of Genetics* 38:645–679.

Pamilo, P. and M. Nei. 1988. "Relationships between Gene Trees and Species Trees." *Molecular Biology and Evolution* 5(5):568–583.

Parducci, L., Y. Suyama, M. Lascoux, and K. D. Bennett. 2005. "Ancient DNA from Pollen: A Genetic Record of Population History in Scots Pine." *Molecular Ecology* 14: 2873–2882.

Pawley, A. and R. C. Green. 1973. "Dating the Dispersal of the Oceanic Languages." *Oceanic Linguistics* 12:1–67.

Pedrosa, S., M. Uzun, J-J. Arranz, B. Gutiérrez-Gill, F. S. Primitivo, and Y. Bayón. 2005. "Evidence of Three Maternal Lineages in Near Eastern Sheep Supporting Multiple Domestication Events." *Proceedings of the Royal Society B: Biological Sciences* 272(1577):2211–2217.

Perego, U. A., A. Achilli, N. Angerhofer, M. Accetturo, M. Pala, A. Olivieri, B. H. Kashani, K. H. Ritchie, R. Scozzari, Q-P. Kong, and others. 2009. "Distinctive Paleo-Indian Migration Routes from Beringia Marked by Two Rare mtDNA Haplogroups." *Current Biology* 19(1):1–8.

Pereira, F., S. Queirós, L. Gusmão, I. J. Nijman, E. Cuppen, J. A. Lenstra, Econogene Consortium, S. J. M. Davis, F. Nejmeddine, and A. Amorim. 2009. "Tracing the History of Goat Pastoralism: New Clues from Mitochondrial and Y Chromosome DNA in North Africa." *Molecular Biology and Evolution* 26(12):2765–2773.

Pereira, L., V. Černý, M. Cerezo, N. M. Silva, M. Hájek, A. Vašíková, M. Kujanová, R. Brdička, and A. Salas. 2010. "Linking the Sub-Saharan and West Eurasian Gene Pools: Maternal and Paternal Heritage of the Tuareg Nomads of the African Sahel." *European Journal of Human Genetics* 18:915–923.

Pereira, L., M. Richards, A. Goios, A. Alonso, C. Albarrán, O. Garcia, D. M. Behar, M. Gölge, J. Hatina, A. L. Gazal, and others. 2005. "High-resolution mtDNA Evidence for the Late-glacial Resettlement of Europe from an Iberian Refugium." *Genome Research* 15(1):19–24.

Perry, G. H., N. J. Dominy, K. G. Claw, A. S. Lee, H. Fiegler, R. Redon, J. Werner, F. A. Villanea, J. L. Mountain, R. Misra, and others. 2007. "Diet and the Evolution of Human Amylase Gene Copy Number Variation." *Nature Genetics* 39(10):1256–1260.

Peters, J., D. Helmer, A. Von Den Driesch, and M. Saña Segui. 1999. "Early Animal Husbandry in the Northern Levant." *Paléorient* 25:27–48.

Phillipson, D. W. 2005. *African Archaeology*. Cambridge: Cambridge University Press.

Pier, G. B., M. Grout, T. Zaidi, G. Meluleni, S. S. Mueschenborn, G. Banting, R. Ratcliff, M. J. Evans, and W. H. Colledge. 1998. "*Salmonella typhi* Uses CFTR to Enter Intestinal Epithelial Cells." *Nature* 393(6680):79–82.

Pierson, M. J., R. Martinez-Arias, B. R. Holland, N. J. Gemmell, M. E. Hurles, and D. Penny. 2006. "Deciphering Past Human Population Movements in Oceania: Provably Optimal Trees of 127 mtDNA Genomes." *Molecular Biology and Evolution* 23(10):1966–1975.

Piperno, D. R. 1993. "Phytolith and Charcoal Records from Deep Lake Cores in the American Tropics." In: D. M. Pearsall and D. R. Piperno, editors. *Current Research in Phytolith Analysis: Applications in Archaeology and Paleoecology*. Philadelphia, PA: The University Museum of Archaeology and Anthropology, University of Pennsylvannia. pp. 58–71.

Piperno, D. R., M. B. Bush, and P. A. Colinvaux. 1991. "Paleoecological Perspectives on Human Adaptation in Central Panama. II the Holocene." *Geoarchaeology* 6(3):227–250.

Poinar, H. N. and B. A. Stankiewicz. 1999. "Protein Preservation and DNA Retrieval from Ancient Tissues." *Proceedings of the National Academy of Sciences of the United States of America* 96(15):8426–8431.

Poinar, H. N., M. Höss, J. L. Bada, and S. Pääbo. 1996. "Amino Acid Racemization and the Preservation of Ancient DNA." *Science* 272:864–866.

Poinar, H. N., C. Schwarz, J. Qi, B. Shapiro, R. D. E. MacPhee, B. Buigues, A. Tikhonov, D. H. Huson, L. P. Tomsho, A. Auch, and others. 2006. "Metagenomics to Paleogenomics: Large-scale Sequencing of Mammoth DNA." *Science* 311(5759):392–394.

Poulakakis, N., A. Tselikas, I. Bitsakis, M. Mylonas, and P. Lymberakis. 2007. "Ancient DNA and the Genetic Signature of Ancient Greek Manuscripts." *Journal of Archaeological Science* 34(5):675–680.

Quintana-Murci, L., H. Quach, C. Harmant, F. Luca, B. Massonnet, E. Patin, L. Sica, P. Mouguiama-Daouda, D. Comas, S. Tzur, and others. 2008. "Maternal Traces of Deep Common Ancestry and Asymmetric Gene Flow between Pygmy Hunter-gatherers and Bantu-speaking Farmers." *Proceedings of the National Academy of Sciences of the United States of America* 105(5):1596–1601.

Raoult, D., G. Aboudharam, E. Crubézy, G. Larrouy, B. Ludes, and M. Drancourt. 2000. "Molecular Identification by 'Suicide PCR' of *Yersinia pestis* as the Agent of Medieval Black Death." *Proceedings of the National Academy of Sciences of the United States of America* 97(23):12800–12803.

Rasmussen, M., X. Guo, Y. Wang, K. E. Lohmueller, S. Rasmussen, A. Albrechtsen, L. Skotte, S. Lindgreen, M. Metspalu, T. Jombart, and others. 2011. "An Aboriginal Australian Genome Reveals Separate Human Dispersals into Asia." *Science* 333(6052):94–98.

Rasmussen, M., Y. R. Li, S. Lindgreen, J. S. Pedersen, A. Albrechtsen, I. Moltke, M. Metspalu, E. Metspalu, T. Kivisild, R. Gupta, and others. 2010. "Ancient Human Genome Sequence of an Extinct Palaeo-Eskimo. *Nature* 463(7282):757–762.

Razafindrazaka, H., F-X. Ricaut, M. P. Cox, M. Mormina, J-M. Dugoujon, L. P. Randriamarolaza, E. Guitard, L. Tonasso, B. Ludes, and E. Crubézy. 2010. "Complete > Mitochondrial DNA Sequences Provide New Insights into the Polynesian Motif and the Peopling of Madagascar." *European Journal of Human Genetics* 18(5):575–581.

Reardon, J. 2001. "The Human Genome Diversity Project: A Case Study in Coproduction." *Social Studies of Science* 31(3):357–388.

Redd, A. J., N. Takezaki, S. T. Sherry, S. T. McGarvey, A. S. Sofro, and M. Stoneking. 1995. "Evolutionary History of the COII/tRNA(Lys) Intergenic 9-base-pair Deletion in Human Mitochondrial DNAs from the Pacific." *Molecular Biology and Evolution* 12(4):604–615.

Reed, F. A., E. J. Kontanis, K. A. R. Kennedy, and C. F. Aquadro. 2003. "Brief Communication: Ancient DNA Prospects from Sri Lankan Highland Dry Caves Support an Emerging Global Pattern." *American Journal of Physical Anthropology* 121(2):112–116.

Reich, D., R. E. Green, M. Kircher, J. Krause, N. Patterson, E. Y. Durand, B. Viola, A. W. Briggs, U. Stenzel, P. L. F. Johnson, and others. 2010. "Genetic History of an Archaic Hominin Group from Denisova Cave in Siberia." *Nature* 468(7327):1053–1060.

Reich, D., N. Patterson, M. Kircher, F. Delfin, M. Nandineni, I. Pugach, A.M-S. Ko, Y-C. Ko, T. Jinam, M. Phipps, and others. 2011. "Denisova Admixture and the First Modern

Human Dispersals into Southeast Asia and Oceania." *The American Journal of Human Genetics* 89(4):516–528.

Reich, D., K. Thangaraj, N. Patterson, A. L. Price, and L. Singh. 2009. "Reconstructing Indian Population History." *Nature* 461(7263):489–494.

Reidla, M., T. Kivisild, E. Metspalu, K. Kaldma, K. Tambets, H-V. Tolk, J. Parik, E-L. Loogväli, M. Derenko, B. Malyarchuk, and others. 2003. "Origin and Diffusion of mtDNA Haplogroup X." *American Journal of Human Genetics* 73(5):1178–1190.

Renfrew, C. 2010. "Archaeogenetics—Towards a 'New Synthesis'?" *Current Biology* 20(4):R162–R165.

Richards, M. and Y. Macaulay. 2001. "The Mitochondrial Gene Tree Comes of Age." *American Journal of Human Genetics* 68(6):1315–1320.

Richards, M., H. Côrte-Real, P. Forster, V. Macaulay, H. Wilkinson-Herbots, A. Demaine, S. Papiha, R. Hedges, H. J. Bandelt, and B. Sykes. 1996. "Paleolithic and Neolithic Lineages in the European Mitochondrial Gene Pool." *American Journal of Human Genetics* 59(1):185–203.

Rick, T. C., J. M. Erlandson, R. L. Vellanoweth, T. J. Braje, P. W. Collins, D. A. Guthrie, and T. W. Stafford, Jr. 2009. "Origins and Antiquity of the Island Fox (*Urocyon littoralis*) on California's Channel Islands." *Quaternary Research* 71(2):93–98.

Ristaino, J. B. 2002. "Tracking Historic Migrations of the Irish Potato Famine Pathogen, *Phytophthora infestans*." *Microbes and Infection* 4(13):1369–1377.

Ristaino, J. B., C. T. Groves, and G. R. Parra. 2001. "PCR Amplification of the Irish Potato Famine Pathogen from Historic Specimens." *Nature* 411(6838):695–697.

Roberts, C. and S. Ingham. 2008. "Using Ancient DNA Analysis in Palaeopathology: A Critical Analysis of Published Papers, with Recommendations for Future Work." *International Journal of Osteoarchaeology* 18(6):600–613.

Roberts, L. 1992a. "How to Sample the World's Genetic Diversity." *Science* 257(5074):1204–1205.

Roberts, L. 1992b. "Genome Diversity Project: Anthropologists Climb (gingerly) on Board." *Science* 258(5086):1300–1301.

Robins, J. H., E. Matisoo-Smith, and L. Furey. 2001. "Hit or Miss? Factors Affecting DNA Preservation in Pacific Archaeological Material." In: M. Jones and P. Sheppard, editors. *Australian Connections and New Directions: Proceedings of the 7th Australasian Archaeometry Conference. Research in Anthropology and Linguistics* No. 5. Auckland: Department of Anthropology, University of Auckland.

Robins, J. H., P. A. McLenachan, M. J. Phillips, L. Craig, H. A. Ross, and E. Matisoo-Smith. 2008. "Dating of Divergences within the *Rattus* Genus Phylogeny Using Whole Mitochondrial Genomes." *Molecular Phylogenetics and Evolution* 49(2):460–466.

Røed, K. H., Ø Flagstad, M. Nieminen, Ø Holand, M. J. Dwyer, N. Røv, and C. Vilà. 2008. "Genetic Analyses Reveal Independent Domestication Origins of Eurasian Reindeer." *Proceedings of the Royal Society B: Biological Sciences* 275(1645):1849–1855.

Rogaev, E. I., A. P. Grigorenko, Y. K. Moliaka, G. Faskhutdinova, A. Goltsov, A. Lahti, C. Hildebrandt, E. L. W. Kittler, and I. Morozova. 2009. "Genomic Identification in the Historical Case of the Nicholas II Royal Family." *Proceedings of the National Academy of Sciences of the United States of America* 106(13):5258–5263.

Rohde, D. L.T., S. Olson, and J. T. Chang. 2004. "Modelling the Recent Common Ancestry of all Living Humans." *Nature* 431(7008):562–566.

Rohland, N., H. Siedel, and M. Hofreiter. 2004. "Nondestructive DNA Extraction Method for Mitochondrial DNA Analyses of Museum Specimens." *BioTechniques* 36(5):814.

Rollo, F., L. Ermini, S. Luciani, I. Marota, C. Olivieri, and D. Luiselli. 2006. "Fine Characterisation of the Iceman's mtDNA Haplogroup." *American Journal of Physical Anthropology* 130(4):557–564.

Rollo, F., M. Ubaldi, L. Ermini, and I. Marota. 2002. "Ötzi's Last Meals: DNA Analysis of the Intestinal Contents of the Neolithic Glacier Mummy from the Alps." *Proceedings of the National Academy of Sciences of the United States of America* 99(20):12594–12599.

Rootsi, S., T. Kivisild, G. Benuzzi, H. Help, M. Bermisheva, I. Kutuev, L. Barać, M. Peričić, O. Balanovsky, A. Pshenichnov, and others. 2004. "Phylogeography of Y-chromosome Haplogroup I Reveals Distinct Domains of Prehistoric Gene Flow in Europe." *American Journal of Human Genetics* 75(1):128–137.

Rootsi, S., L. A. Zhivotovsky, M. Baldovič, M. Kayser, I. A. Kutuev, R. Khusainova, M. A. Bermisheva, M. Gubina, S. A. Fedorova, A. M. Ilumäe, and others. 2007. "A Counter-clockwise Northern Route of the Y-chromosome Haplogroup N from Southeast Asia towards Europe." *European Journal of Human Genetics* 15(2):204–211.

Rosa, A., C. Ornelas, M. A. Jobling, A. Brehm, and R. Villems. 2007. "Y-chromosomal Diversity in the Population of Guinea-Bissau: A Multiethnic Perspective." *BMC Evolutionary Biology* 7:124.

Rose, J. I. 2010. "New Light on Human Prehistory in the Arabo-Persian Gulf Oasis." *Current Anthropology* 51(6):849–883.

Rothberg, J. M. and J. H. Leamon. 2008. "The Development and Impact of 454 Sequencing." *Nature Biotechnology* 26(10):1117–1124.

Ruvolo, M., T. R. Disotell, M. W. Allard, W. M. Brown, and R. L. Honeycutt. 1991. "Resolution of the African Hominoid Trichotomy by Use of a Mitochondrial Gene Sequence." *Proceedings of the National Academy of Sciences of the United States of America* 88(4):1570–1574.

Ruvolo, M., S. Zehr, M. Vondornum, D. Pan, B. Chang, and J. Lin. 1993. "Mitochondrial COII Sequences and Modern Human Origins." *Molecular Biology and Evolution* 10(6):1115–1135.

Sablin, M. V. and G. A. Khlopachev. 2002. "The Earliest Ice Age Dogs: Evidence from Eliseevichi I." *Current Anthropology* 43(5):795–799.

Salas, A., M. Richards, T. De la Fe, M-V. Lareu, B. Sobrino, P. Sánchez-Diz, V. Macaulay, and Á. Carracedo. 2002. "The Making of the African mtDNA Landscape." *American Journal of Human Genetics* 71(5):1082–1111.

Sallares, R. and S. Gomzi. 2001. "Biomolecular Archaeology of Malaria." *Ancient Biomolecules* 3:195–213.

Sampietro, M. L., M. T. P. Gilbert, O. Lao, D. Caramelli, M. Lari, J. Bertranpetit, and C. Lalueza-Fox. 2006. "Tracking Down Human Contamination in Ancient Human Teeth." *Molecular Biology and Evolution* 23(9):1801–1807.

Sang, T. and S. Ge. 2007. "Genetics and Phylogenetics of Rice Domestication." *Current Opinion in Genetics and Development* 17(6):533–538.

Santos, F. R., A. Pandya, C. Tyler-Smith, S. D. J. Pena, M. Schanfield, W. R. Leonard, L. Osipova, M. H. Crawford, and R. J. Mitchell. 1999. "The Central Siberian Origin for Native American Y Chromosomes." *American Journal of Human Genetics* 64(2):619–628.

Sarich, V. M. and J. Cronin. 1976. "Molecular Systematics of the Primates." In: M. Goodman and R. Tashian, editors. *Molecular Anthropology: Genes and Proteins in the Evolutionary Aspect of the Primates.* New York: Plenum. pp. 141–171.

Sarich, V. M. 1977. "Albumin Phylogenetics." In: V. W. Rosenoer, M. Oratz, and M. A. Rothschild, editors. *Albumin Structure, Function and Uses.* New York: Pergamon Press. pp. 85–111.

Sarich, V. M. and A. C. Wilson. 1967a. "Immunological Time Scale for Hominid Evolution." *Science* 158(3805):1200–1203.

Sarich, V. M. and A. C. Wilson. 1967b. "Rates of Albumin Evolution in Primates." *Proceedings of the National Academy of Sciences of the United States of America* 58(1):142–148.

Savolainen, P., T. Leitner, A. N. Wilton, E. Matisoo-Smith, and J. Lundeberg. 2004. "A Detailed Picture of the Origin of the Australian Dingo, Obtained from the Study of Mitochondrial DNA." *Proceedings of the National Academy of Sciences of the United States of America* 101(33):12387–12390.

Savolainen, P., Y-P. Zhang, J. Luo, L. Lundeberg, and T. Leitner. 2002. "Genetic Evidence for an East Asian Origin of Domestic Dogs." *Science* 298(5598):1610–1613.

Scheinfeldt, L., F. Friedlaender, J. Friedlaender, K. Latham, G. Koki, T. Karafet, M. Hammer, and J. Lorenz. 2006. "Unexpected NRY Chromosome Variation in Northern Island Melanesia." *Molecular Biology and Evolution* 23(8):1628–1641.

Schlumbaum, A., P. F. Campos, S. Volken, M. Volken, A. Hafner, and J. Schibler. 2010. "Ancient DNA, a Neolithic Legging from the Swiss Alps and the History of Early Goat." *Journal of Archaeological Science* 37(6):1247–1251.

Schmidt, D., S. Hummel, and B. Herrmann. 2003. "Brief Communication: Multiplex X/Y-PCR Improves Sex Identification in aDNA Analysis." *American Journal of Physical Anthropology* 121(4):337–341.

Schneider, S. and L. Excoffier. 1999. "Estimation of Past Demographic Parameters from the Distribution of Pairwise Differences When the Mutation Rates Vary among Sites: Application to Human Mitochondrial DNA." *Genetics* 152(3):1079–1089.

Schuenemann, V. J., K. Bos, S. DeWitte, S. Schmedes, J. Jamieson, A. Mittnik, S. Forrest, B. K. Coombes, J. W. Wood, D. J. D. Earn, and others. 2011. "Targeted Enrichment of Ancient Pathogens Yielding the pPCP1 Plasmid of *Yersinia pestis* from Victims of the Black Death." *Proceedings of the National Academy of Sciences of the United States of America* 108(38):E746–E752.

Schurr, T. G. and S. T. Sherry. 2004. "Mitochondrial and Y Chromosome Diversity and the Peopling of the Americas: Evolutionary and Demographic Evidence." *American Journal of Human Biology* 16(4):420–439.

Schurr, T. G., A. C. Owings, J. B. Gaieski, I. Kritsch, A. Andre, C. Lennie, H. Zillges, K. Keating, and The Genographic C. 2011. "Genetic Histories of Gwich'in and Inuvialuit Populations of Northwest Territories, Canada." *American Journal of Physical Anthropology* 144:266–267.

Seal, U. S., N. I. Phillips, and A. W. Erickson. 1970. "Carnivora Systematics: Immunological Relationships of Bear Albumins." *Comparative Biochemistry and Physiology* 32(1):33–48.

Searle, J. B., C. S. Jones, I. Gündüz, M. Scascitelli, E. P. Jones, J. S. Herman, R. V. Rambau, L. R. Noble, R. J. Berry, M. D. Giménez, and others. 2009. "Of Mice and (Viking?) Men: Phylogeography of British and Irish House Mice." *Proceedings of the Royal Society B: Biological Sciences* 276(1655):201–207.

Seelenfreund, D., A. Clarke, N. Oyanedel, R. Piña, S. Lobos, E. Matisoo-Smith, and A. Seelenfreund. 2010. "Paper Mulberry (*Broussonetia papyrifera*) as a Commensal Model for Human Mobility in Oceania: Anthropological, Botanical and Genetic Considerations." *New Zealand Journal of Botany* 48(3):231–247.

Seidler, H., W. Bernhard, M. Teschler-Nicola, W. Platzer, D. zur Nedden, R. Henn, A. Oberhauser, and T. Sjovold. 1992. "Some Anthropological Aspects of the Prehistoric Tyrolean Ice Man." *Science* 258 (5081):455–457.

Semino, O., C. Magri, G. Benuzzi, A. A. Lin, N. Al-Zahery, V. Battaglia, L. Maccioni, C. Triantaphyllidis, P. Shen, P. J. Oefner, and others. 2004. "Origin, Diffusion, and Differentiation of Y-chromosome Haplogroups E and J: Inferences on the Neolithization of Europe and Later Migratory Events in the Mediterranean Area." *American Journal of Human Genetics* 74(5):1023–1034.

Serjeantson, S. W. and A. V. S. Hill, editors. 1989. *The Colonisation of the Pacific: A Genetic Trail.* Oxford: Clarendon Press.

Serre, D., A. Langaney, M. Chech, M. Teschler-Nicola, M. Paunovic, P. Mennecier, M. Hofreiter, G. Possnert, and S. Pääbo. 2004. "No Evidence of Neandertal mtDNA Contribution to Early Modern Humans." *PLoS Biology* 2(3):313–317.

Service, R. F. 2006. "The Race for the $1000 Genome." *Science* 311(5767):1544–1546.

Shackleton, D. M. 1997. *Wild Sheep and Goats and their Relatives: Status Survey and Conservation Action Plan for Caprinae.* Gland, Switzerland: IUCN.

Shapiro, B., A. J. Drummond, A. Rambaut, M. C. Wilson, P. E. Matheus, A. V. Sher, O. G. Pybus, M. T. P. Gilbert, I. Barnes, J. Binladen, and others. 2004. "Rise and Fall of the Beringian Steppe Bison." *Science* 306(5701):1561–1565.

Shi, H., H. Zhong, Y. Peng, Y-L. Dong, X-B. Qi, F. Zhang, L-F. Liu, S-J. Tan, R. Z. Ma, C-J. Xiao, and others. 2008. "Y Chromosome Evidence of Earliest Modern Human Settlement in East Asia and Multiple Origins of Tibetan and Japanese Populations." *BMC Biology* 6:45.

Shi, W., Q. Ayub, M. Vermeulen, R-G. Shao, S. Zuniga, K. van der Gaag, P. de Knijff, M. Kayser, Y. Xue, and C. Tyler-Smith. 2010. "A Worldwide Survey of Human Male Demographic History Based on Y-SNP and Y-STR Data from the HGDP CEPH Populations." *Molecular Biology and Evolution* 27(2):385–393.

Sibley, C. G. and J. E. Ahlquist. 1984. "The Phylogeny of the Hominoid Primates, as Indicated by DNA-DNA Hybridization." *Journal of Molecular Evolution* 20(1):2–15.

Sigurðardóttir, S., A. Helgason, J. R. Gulcher, K. Stefánsson, and P. Donnelly. 2000. "The Mutation Rate in the Human mtDNA Control Region." *American Journal of Human Genetics* 66(5):1599–1609.

Slate, J. and N. J. Gemmell. 2004. "Eve 'n' Steve: Recombination of Human Mitochondrial DNA." *Trends in Ecology and Evolution* 19(11):561–563.

Smith, A. B. 2005. *African Herders: Emergence of Pastoral Traditions*. Walnut Creek: AltaMira.

Smith, C. I., A. T. Chamberlain, M. S. Riley, C. Stringer, and M. J. Collins. 2003. "The Thermal History of Human Fossils and the Likelihood of Successful DNA Amplification." *Journal of Human Evolution* 45(3):203–217.

Smith, C. I., O. E. Craig, R. V. Prigodich, C. M. Nielsen–Marsh, M. M. E. Jans, C. Vermeer, and M. J. Collins. 2005. "Diagenesis and Survival of Osteocalcin in Archaeological Bone." *Journal of Archaeological Science* 32(1):105–113.

Soares, P., A. Achilli, O. Semino, W. Davies, V. Macaulay, H-J. Bandelt, A. Torroni, and M. B. Richards. 2010. "The Archaeogenetics of Europe." *Current Biology* 20(4):R174–R183.

Soares, P., T. Rito, J. Trejaut, M. Mormina, C. Hill, E. Tinkler-Hundal, M. Braid, D. J. Clarke, J-H. Loo, N. Thomson, and others. 2011. "Ancient Voyaging and Polynesian Origins." *American Journal of Human Genetics* 88(2):239–247.

Sokal, R. R. and C. D. Michener. 1958. "A Statistical Method for Evaluating Systematic Relationships." *University of Kansas Science Bulletin* 38: 1409–1438.

Speller, C. F., B. M. Kemp, S. D. Wyatt, C. Monroe, W. D. Lipe, U. M. Arndt, and D. Y. Yang. 2010. "Ancient Mitochondrial DNA Analysis Reveals Complexity of Indigenous North American Turkey Domestication." *Proceedings of the National Academy of Sciences of the United States of America* 107(7):2807–2812.

Steadman, D. W., G. K. Pregill, and D. V. Burley. 2002. "Rapid Prehistoric Extinction of Iguanas and Birds in Polynesia." *Proceedings of the National Academy of Sciences of the United States of America* 99(6):3673–3677.

Stevens, J., A. L. Corper, C. F. Basler, J. K. Taubenberger, P. Palese, and I. A. Wilson. 2004. "Structure of the Uncleaved Human H1 Hemagglutinin from the Extinct 1918 Influenza Virus." *Science* 303:1866–1870.

Stiller, M., R. E. Green, M. Ronan, J. F. Simons, L. Du, W. He, M. Egholm, J. M. Rothberg, S. G. Keats, N. D. Ovodov, and others. 2006. "Patterns of Nucleotide Misincorporations during Enzymatic Amplification and Direct Large-scale Sequencing of Ancient DNA." *Proceedings of the National Academy of Sciences of the United States of America* 103(37):13578–13584.

Stokstad, E. 2003. "Ancient DNA Pulled from Soil." *Science* 300:407.

Stone, A. C., G. R. Milner, S. Pääbo, and M. Stoneking. 1996. "Sex Determination of Ancient Human Skeletons Using DNA." *American Journal of Physical Anthropology* 99(2):231–238.

Stoneking, M. and F. Delfin. 2010. "The Human Genetic History of East Asia: Weaving a Complex Tapestry." *Current Biology* 20(4):R188–R193.

Storey, A. A., T. Ladefoged, and E. A. Matisoo-Smith. 2008. "Counting Your Chickens: Density and Distribution of Chicken Remains in Archaeological Sites of Oceania." *International Journal of Osteoarchaeology* 18(3):240–261.

Storey, A. A., D. Quiroz, J. M. Ramírez, N. Beavan-Athfield, D. J. Addison, R. Walter, T. Hunt, J. S. Athens, L. Huynen, and E. A. Matisoo-Smith. 2008. "Pre-Colombian Chickens, Dates, Isotopes, and mtDNA." *Proceedings of the National Academy of Sciences of the United States of America* 105(48):E99.

Storey, A. A., J. M. Ramírez, D. Quiroz, D. V. Burley, D. J. Addison, R. Walter, A. J. Anderson, T. L. Hunt, J. S. Athens, L. Huynen, and others. 2007. "Radiocarbon and DNA Evidence for a Pre-Columbian Introduction of Polynesian Chickens to Chile." *Proceedings of the National Academy of Sciences of the United States of America* 104(25):10335–10339.

Storey, A. A., M. Spriggs, S. Bedford, S. C. Hawkins, J. H. Robins, L. Huynen, and E. Matisoo-Smith. 2010. "Mitochondrial DNA from 3000-Year Old Chickens at the Teouma Site, Vanuatu." *Journal of Archaeological Science* 37(10):2459–2468.

Stringer, C. B. and C. Gamble. 1993. *In Search of the Neanderthals: Solving the Puzzle of Modern Human Origins*. New York: Thames & Hudson.

Su, B., L. Jin, P. Underhill, J. Martinson, N. Saha, S. T. McGarvey, M. D. Shriver, J. Y. Chu, P. Oefner, R. Chakraborty, and others. 2000. "Polynesian Origins: Insights from the Y Chromosome." *Proceedings of the National Academy of Sciences of the United States of America* 97(15):8225–8228.

Suchentrunk, F., H. Ben Slimen, C. Stamatis, H. Sert, M. Scandura, M. Apollonio, and Z. Mamuris. 2006. "Molecular Approaches Revealing Prehistoric, Historic, or Recent Translocations and Introductions of Hares (Genus *Lepus*) by Humans." *Human Evolution* 21(2):151–165.

Sugimoto, C., M. Hasegawa, H-Y. Zheng, V. Demenev, Y. Sekino, K. Kojima, T. Honjo, H. Kida, T. Hovi, T. Vesikari, and others. 2002. "JC Virus Strains Indigenous to Northeastern Siberians and Canadian Inuits are Unique but Evolutionally Related to Those Distributed throughout Europe and Mediterranean Areas." *Journal of Molecular Evolution* 55(3):322–335.

Summerhayes, G. R., M. Leavesley, A. Fairbairn, H. Mandui, J. Field, A. Ford, and R. Fullagar. 2010. "Human Adaptation and Plant Use in Highland New Guinea 49,000 to 44,000 Years Ago." *Science* 330(6000):78–81.

Svensson, E. M., C. Anderung, J. Baubliene, P. Persson, H. Malmström, C. Smith, M. Vretemark, L. Daugnora, and A. Götherström. 2007. "Tracing Genetic Change over Time Using Nuclear SNPs in Ancient and Modern Cattle." *Animal Genetics* 38(4):378–383.

Sweeney, M. and S. McCouch. 2007. "The Complex History of the Domestication of Rice." *Annals of Botany* 100(5):951–957.

Swofford, D. L. 1991. "PAUP: Phylogenetic Analysis Using Parsimony, Version 3.1." Computer program distributed by the Illinois Natural History Survey, Champaign, Illinois.

Sykes, B., A. Leiboff, J. Lowbeer, S. Tetzner, and M. Richards. 1995. "The Origins of the Polynesians—An Interpretation from Mitochondrial Lineage Analysis." *American Journal of Human Genetics* 57(6):1463–1475.

Tabbada, K. A., J. Trejaut, J-H. Loo, Y-M. Chen, M. Lin, M. Mirazón-Lahr, T. Kivisild, and M. C. A. De Ungria. 2010. "Philippine Mitochondrial DNA Diversity: A Populated Viaduct between Taiwan and Indonesia?" *Molecular Biology and Evolution* 27(1):21–31.

Takasaka, T., J. J. Miranda, C. Sugimoto, R. Paraguison, H-Y. Zheng, T. Kitamura, and Y. Yogo. 2004. "Genotypes of JC Virus in Southeast Asia and the Western Pacific:

Implications for Human Migrations from Asia to the Pacific." *Anthropological Science* 112(1):53–59.

Tambets, K., S. Rootsi, T. Kivisild, H. Help, P. Serk, E-L. Loogväli, H-V. Tolk, M. Reidla, E. Metspalu, L. Pliss, and others. 2004. "The Western and Eastern Roots of the Saami—The Story of Genetic 'Outliers' Told by Mitochondrial DNA and Y Chromosomes." *American Journal of Human Genetics* 74(4):661–682.

Tamm, E., T. Kivisild, M. Reidla, M. Metspalu, D. G. Smith, C. J. Mulligan, C. M. Bravi, O. Rickards, C. Martinez-Labarga, E. K. Khusnutdinova, and others. 2007. "Beringian Standstill and Spread of Native American Founders." *PLoS One* 2(9):e829.

Tang, H., D. O. Siegmund, P. Shen, P. J. Oefner, and M. W. Feldman. 2002. "Frequentist Estimation of Coalescence Times from Nucleotide Sequence Data Using a Tree-based Partition." *Genetics* 161:447–459.

Tang, T. and S. Shi. 2007. "Molecular Population Genetics of Rice Domestication." *Journal of Integrative Plant Biology* 49(6):769–775.

Tang, X., G. Zhao, and L. Ping. 2011. "Wood Identification with PCR Targeting Noncoding Chloroplast DNA." *Plant Molecular Biology* 77(6):609–617.

Tapio, M., N. Marzanov, M. Ozerov, M. Inkulov, G. Gonzarenko, T. Kiselyova, M. Murawski, H. Viinalass, and J. Kantanen. 2006. "Sheep Mitochondrial DNA Variation in European, Caucasian, and Central Asian Areas." *Molecular Biology and Evolution* 23(9):1776–1783.

Tattersall, I. and J. H. Schwartz. 1999. "Hominids and Hybrids: The Place of Neanderthals in Human Evolution." *Proceedings of the National Academy of Sciences of the United States of America* 96(13):7117–7119.

Taubenberger, J. K., A. H. Reid, A. E. Krafft, K. E. Bijwaard, and T. G. Fanning. 1997. "Initial Genetic Characterization of the 1918 'Spanish' Influenza Virus." *Science* 275:1793–1796.

Taylor, G. M., M. Crossey, J. Saldanha, and T. Waldron. 1996. "DNA from *Mycobacterium tuberculosis* Identified in Mediaeval Human Skeletal Remains Using Polymerase Chain Reaction." *Journal of Archaeological Science* 23(5):789–798.

Taylor, G. M., S. A. Mays, and J. F. Huggett. 2009. "Ancient DNA (aDNA) Studies of Man and Microbes: General Similarities, Specific Differences." *International Journal of Osteoarchaeology* 20(6):747–751.

Taylor, G. M., R. Rutland, and T. Molleson. 1997. "A Sensitive Polymerase Chain Reaction Method for the Detection of *Plasmodium* Species DNA in Ancient Human Remains." *Ancient Biomolecules* 1:193–203.

Taylor, G. M., C. L. Watson, A. S. Bouwman, D. N. J. Lockwood, and S. A. Mays. 2006. "Variable Nucleotide Tandem Repeat (VNTR) Typing of Two Palaeopathological Cases of Lepromatous Leprosy from Mediaeval England." *Journal of Archaeological Science* 33(11):1569–1579.

Taylor, G. M., S. Widdison, I. N. Brown, and D. Young. 2000. "A Mediaeval Case of Lepromatous Leprosy from 13–14th Century Orkney, Scotland." *Journal of Archaeological Science* 27(12):1133–1138.

Taylor, G. M., D. B. Young, and S. A. Mays. 2005. "Genotypic Analysis of the Earliest Known Prehistoric Case of Tuberculosis in Britain." *Journal of Clinical Microbiology* 43(5):2236–2240.

Templeton, A. R. 1991. "Human Origins and Analysis of Mitochondrial DNA Sequences." *Science* 255(5045):737–739.

Terauchi, R., V. A. Chikaleke, G. Thottappilly, and S. K. Hahn. 1992. "Origin and Phylogeny of Guinea Yams as Revealed by RFLP Analysis of Chloroplast DNA and Nuclear Ribosomal DNA." *Theoretical and Applied Genetics* 83(6–7):743–751.

Thangaraj, K., G. Chaubey, V. K. Singh, A. Vanniarajan, I. Thanseem, A. G. Reddy, and L. Singh. 2006. "In Situ Origin of Deep Rooting Lineages of Mitochondrial Macrohaplogroup 'M' in India." *BMC Genomics* 7:151.

Thangaraj, K., L. Singh, A. G. Reddy, V. R. Rao, S. C. Sehgal, P. A. Underhill, M. Pierson, I. G. Frame, and E. Hagelberg. 2003. "Genetic Affinities of the Andaman Islanders, a Vanishing Human Population." *Current Biology* 13(2):86–93.

The 1000 Genome Consortium. 2010. "A Map of Human Genome Variation from Population-scale Sequencing." *Nature* 467(7319):1061–1073.

Theilmann, J. and F. Cate. 2007. "A Plague of Plagues: The Problem of Plague Diagnosis in Medieval England." *Journal of Interdisciplinary History* 37(3):371–393.

Thomsen, P. F., S. Elias, M. T. P. Gilbert, J. Haile, K. Munch, and others. 2009. "Non-Destructive Sampling of Ancient Insect DNA." *PLoS ONE* 4(4):e5048.

Thomson, R., J. K. Pritchard, P. Shen, P. J. Oefner, and M. W. Feldman. 2000. "Recent Common Ancestry of Human Y Chromosomes: Evidence from DNA Sequence Data." *Proceedings of the National Academy of Sciences of the United States of America* 97(13):7360–7365.

Thorne, A., R. Grün, G. Mortimer, N. A. Spooner, J. J. Simpson, M. McCulloch, L. Taylor, and D. Curnoe. 1999. "Australia's Oldest Human Remains: Age of the Lake Mungo 3 Skeleton." *Journal of Human Evolution* 36(6):591–612.

Thorne, A. G. and M. H. Wolpoff. 1992. "The Multiregional Evolution of Humans." *Scientific American* 266(4):76–79, 82–83.

Tian, C., R. Kosoy, A. Lee, M. Ransom, J. W. Belmont, P. K. Gregersen, and M. F. Seldin. 2008. "Analysis of East Asia Genetic Substructure Using Genome-wide SNP Arrays." *PLoS One* 3(12):e3862.

Tishkoff, S. A., E. Dietzsch, W. Speed, A. J. Pakstis, J. R. Kidd, K. Cheung, B. Bonné-Tamir, A. S. Santachiara-Benerecetti, P. Moral, M. Krings, and others. 1996. "Global Patterns of Linkage Disequilibrium at the CD4 Locus and Modern Human Origins." *Science* 271:1380–1387.

Tishkoff, S. A., M. K. Gonder, B. M. Henn, H. Mortensen, A. Knight, C. Gignoux, N. Fernandopulle, G. Lema, T. B. Nyambo, U. Ramakrishnan, and others. 2007. "History of Click-speaking Populations of Africa Inferred from mtDNA and Y Chromosome Genetic Variation." *Molecular Biology and Evolution* 24(10):2180–2195.

Tishkoff, S. A., F. A. Reed, F. R. Friedlaender, C. Ehret, A. Ranciaro, A. Froment, J. B. Hirbo, A. A. Awomoyi, J-M. Bodo, O. Doumbo, and others. 2009. "The Genetic Structure and History of Africans and African Americans." *Science* 324(5930):1035–1044.

Tocheri, M. W., C. M. Orr, S. G. Larson, T. Sutikna, Jatmiko, E. W. Saptomo, R. A. Due, T. Djubiantono, M. J. Morwood, and W. L. Jungers. 2007. "The Primitive Wrist of *Homo floresiensis* and its Implications for Hominin Evolution." *Science* 317(5845):1743–1745.

Tollenaere, C., C. Brouat, J. M. Duplantier, L. Rahalison, S. Rahelinirina, M. Pascal, H. Moné, G. Mouahid, H. Leirs, and J. F. Cosson. 2010. "Phylogeography of the Introduced Species *Rattus rattus* in the Western Indian Ocean, with Special Emphasis on the Colonization History of Madagascar." *Journal of Biogeography* 37(3):398–410.

Trueman, J. W. H. 2001. "Does the Lake Mungo 3 mtDNA Evidence Stand up to Analysis?" *Archaeology in Oceania* 36(3):163–165.

Tsuda, K., Y. Kikkawa, H. Yonekawa, and Y. Tanabe. 1997. "Extensive Interbreeding Occurred among Multiple Matriarchal Ancestors During the Domestication of Dogs: Evidence from Inter- and Intraspecies Polymorphisms in the D-loop Region of Mitochondrial DNA between Dogs and Wolves." *Genes and Genetic Systems* 72(4):229–238.

Uerpmann, H. P. 1987. *The Ancient Distribution of Ungulate Mammals in the Middle East: Fauna and Archaeological Sites in Southwest Asia and Northeast Africa*. Weisbaden: Dr. Ludwig Reichert Vertag.

Underhill, P. A., G. Passarino, A. A. Lin, P. Shen, M. Mirazón Lahr, R. A. Foley, P. J. Oefner, and L. L. Cavalli-Sforza. 2001. "The Phylogenetics of Y Chromosome Binary Haplotypes and the Origins of Modern Human Populations." *Annals of Human Genetics* 65:43–62.

van den Bergh, G. D., H. J. M. Meijer, R. D. Awe, M. J. Morwood, K. Szabó, L. W. van den Hoek Ostende, T. Sutikna, E. W. Saptomo, P. J. Piper, and K. M. Dobney. 2009. "The Liang Bua Faunal Remains: A 95k.yr. Sequence from Flores, East Indonesia." *Journal of Human Evolution* 57(5):527–537.

Vargha-Khadem, F., D. G. Gadian, A. Copp, and M. Mishkin. 2005. "FOXP2 and the Neuroanatomy of Speech and Language." *Nature Reviews Neuroscience* 6, 131–138.

Venter, C. 2005. "Sea of Genes." *New Scientist* 186(2499):21.

Venter, J. C., M. D. Adams, E. W. Myers, P. W. Li, R. J. Mural, G. G. Sutton, H. O. Smith, M. Yandell, C. A. Evans, R. A. Holt, and others. 2001. "The Sequence of the Human Genome." *Science* 291(5507):1304–1351.

Verdu, P., F. Austerlitz, A. Estoup, R. Vitalis, M. Georges, S. Théry, A. Froment, S. Le Bomin, A. Gessain, J-M. Hombert, and others. 2009. "Origins and Genetic Diversity of Pygmy Hunter-Gatherers from Western Central Africa." *Current Biology* 19(4):312–318.

Verginelli, F., C. Capelli, V. Coia, M. Musiani, M. Falchetti, L. Ottini, R. Palmirotta, A. Tagliacozzo, I. De Grossi Mazzorin, R. Mariani-Costantini. 2005. "Mitochondrial DNA from Prehistoric Canids Highlights Relationships between Dogs and South-East European Wolves." *Molecular Biology and Evolution* 22(12):2541–2551.

Vergnaud, G., Y. J. Li, O. Gorge, Y. Cui, Y. Song, D. Zhou, I. Grissa, S. V. Dentovskaya, M. E. Platonov, A. Rakin, and others. 2007. "Analysis of the Three *Yersinia pestis* CRISPR Loci Provides New Tools for Phylogenetic Studies and Possibly for the Investigation of Ancient DNA." *Advances in Experimental Medicine and Biology* 603:327–338.

Vernesi, C., D. Caramelli, S. Carbonell, and B. Chiarelli. 1999. "Molecular Sex Determination of Etruscan Bone Samples (7th–3rd c. BC): A Reliability Study." *Homo* 50(2):118–126.

Vigilant, L., M. Stoneking, H. Harpending, K. Hawkes, and A. C. Wilson. 1991. "African Populations and the Evolution of Human Mitochondrial DNA." *Science* 253:1503–1507.

Vilà, C., J. E. Maldonado, and R. K. Wayne. 1999. "Phylogenetic Relationships, Evolution, and Genetic Diversity of the Domestic Dog." *Journal of Heredity* 90(1):71–77.

Vilà, C., P. Savolainen, J. E. Maldonado, I. R. Amorim, J. E. Rice, R. L. Honeycutt, K. A. Crandall, J. Lundeberg, and R. K. Wayne. 1997. "Multiple and Ancient Origins of the Domestic Dog." *Science* 276(5319):1687–1689.

Volodko, N. V., E. B. Starikovskaya, I. O. Mazunin, N. P. Eltsov, P. V. Naidenko, D. Wallace, and R. Sukernik. 2008. "Mitochondrial Genome Diversity in Arctic Siberians, with Particular Reference to the Evolutionary History of Beringia and Pleistocenic Peopling of the Americas." *American Journal of Human Genetics* 82(5):1084–1100.

von Hunnius, T. E., D. Yang, B. Eng, J. S. Waye, and S. R. Saunders. 2007. "Digging Deeper into the Limits of Ancient DNA Research on Syphilis." *Journal of Archaeological Science* 34(12):2091–2100.

vonHoldt, B. M., J. P. Pollinger, K. E. Lohmueller, E. Han, H. G. Parker, P. Quignon, J. D. Degenhardt, A. R. Boyko, D. A. Earl, A. Auton, and others. 2010. "Genome-wide SNP and Haplotype Analyses Reveal a Rich History Underlying Dog Domestication." *Nature* 464(7290):898–902.

Vuissoz, A., M. Worobey, N. Odegaard, M. Bunce, C. A. Machado, N. Lynnerup, E. E. Peacock, and M. T. P. Gilbert. 2007. "The Survival of PCR-amplifiable DNA in Cow Leather." *Journal of Archaeological Science* 34(5):823–829.

Wang, L., H. Oota, N. Saitou, F. Jin, T. Matsushita, and S. Ueda. 2000. "Genetic Structure of a 2,500-year-old Human Population in China and its Spatiotemporal Changes." *Molecular Biology and Evolution* 17(9):1396–1400.

Wang, S., C. M. Lewis, Jr., M. Jakobsson, S. Ramachandran, N. Ray, G. Bedoya, W. Rojas, M. V. Parra, J. A. Molina, C. Gallo, and others. 2007. "Genetic Variation and Population Structure in Native Americans." *PLoS Genetics* 3(11):e185.

Ward, R. H., B. L. Frazier, K. Dew-Jager, and S. Pääbo. 1991. "Extensive Mitochondrial Diversity within a Single Amerindian Tribe." *Proceedings of the National Academy of Sciences of the United States of America* 88(19):8720–8724.

Ward, R. H., A. Redd, D. Valencia, B. Frazier, and S. Pääbo. 1993. "Genetic and Linguistic Differentiation in the Americas." *Proceedings of the National Academy of Sciences of the United States of America* 90(22):10663–10667.

Waters, M. R. and T. W. Stafford. 2007. "Redefining the Age of Clovis: Implications for the Peopling of the Americas." *Science* 315(5815):1122–1126.

Watson, E., P. Forster, M. Richards, and H-J. Bandelt. 1997. "Mitochondrial Footsteps of Human Expansions in Africa." *The American Journal of Human Genetics* 61(3):691–704.

Watson, J. D. and F. H. C. Crick. 1953a. "Molecular Structure of Nucleic Acids—A Structure for Deoxyribose Nucleic Acid." *Nature* 171(4356):737–738.

Watson, J. D. and F. H. C. Crick. 1953b. "The Structure of DNA." *Cold Spring Harbor Symposia on Quantitative Biology* 18:123–131.

Wayne, R. K., J. A. Leonard, and C. Vilà. 2006. "Genetic Analysis of Dog Domestication." In: M. Zeder, D. G. Bradley, E. Emshwiller, and B. D. Smith, editors. *Documenting Domestication: New Genetic and Archaeological Paradigms*. Berkeley: University of California Press. pp. 279–293.

Wells, S. and T. Schurr. 2009. "Response to Decoding Implications of the Genographic Project." *International Journal of Cultural Property* 16(2):182–187.

Wen, B., H. Li, D. Lu, X. Song, F. Zhang, Y. He, F. Li, Y. Gao, X. Mao, L. Zhang, and others. 2004. "Genetic Evidence Supports Demic Diffusion of Han Culture." *Nature* 431(7006):302–305.

Wen, B., H. Shi, L. Ren, H. Xi, K. Li, W. Zhang, B. Su, S. Si, L. Jin, and C. J. Xiao. 2004. "The Origin of Mosuo People as Revealed by mtDNA and Y Chromosome Variation." *Science in China Series C-Life Sciences* 47(1):1–10.

Wen, B., X. Xie, S. Gao, H. Li, H. Shi, X. Song, T. Qian, C. Xiao, J. Jin, B. Su, and others. 2004. "Analyses of Genetic Structure of Tibeto-Burman Populations Reveals Sex-biased Admixture in Southern Tibeto-Burmans." *American Journal of Human Genetics* 74(5):856–865.

Wendorf, F. and R. Schild, editors. 1980. *Prehistory of the Eastern Sahara*. New York: Academic Press.

Wendorf, F. and R. Schild. 1998. "Nabta Playa and Its Role in Northeastern African Prehistory." *Journal of Anthropological Archaeology* 17(2):97–123.

Westaway, K. E., M. J. Morwood, R. G. Roberts, J. X. Zhao, T. Sutikna, E. W. Saptomo, and W. J. Rink. 2007. "Establishing the Time of Initial Human Occupation of Liang Bua, Western Flores, Indonesia." *Quaternary Geochronology* 2:337–343.

Wetterstrom, W. 2001. "Foraging and Farming in Egypt: The Transition from Hunting and Gathering to Horticulture in the Nile Valley." In: T. Shaw, P. Sinclair, B. Andah, and A. Okpoko, editors. *The Archaeology of Africa: Food, Metals and Towns*. New York: Routledge. pp. 165–226.

White, N. M. 2008. *Archaeology for Dummies*. Hoboken, New Jersey: John Wiley and Sons.

White, T. D., B. Asfaw, D. DeGusta, H. Gilbert, G. D. Richards, G. Suwa, and F. C. Howell. 2003. "Pleistocene *Homo sapiens* from Middle Awash, Ethiopia." *Nature* 423(6941):742–747.

Wickler, S. and M. Spriggs. 1988. "Pleistocene Human Occupation of the Solomon Islands, Melanesia." *Antiquity* 62(237):703–706.

Wiechmann, I. and G. Grupe. 2005. "Detection of *Yersinia pestis* DNA in Two Early Medieval Skeletal Finds from Aschheim (Upper Bavaria, 6th century AD)." *American Journal of Physical Anthropology* 126(1):48–55.

Wilbur, A. K., A. S. Bouwman, A. C. Stone, C. A. Roberts, L. A. Pfister, J. E. Buikstra, and T. A. Brown. 2009. "Deficiencies and Challenges in the Study of Ancient Tuberculosis DNA." *Journal of Archaeological Science* 36(9):1990–1997.

Wilder, J. A., S. B. Kingan, Z. Mobasher, M. M. Pilkington, and M. F. Hammer. 2004. "Global Patterns of Human Mitochondrial DNA and Y-chromosome Structure are not Influenced by Higher Migration Rates of Females Versus Males." *Nature Genetics* 36(10):1122–1125.

Wilkins, M. H. F., A. R. Stokes, and H. R. Wilson. 1953. "Molecular Structure of Deoxypentose Nucleic Acids." *Nature* 4356:738–740.

Willerslev, E., E. Cappellini, W. Boomsma, R. Nielsen, M. B. Hebsgaard, T. B. Brand, M. Hofreiter, M. Bunce, H. N. Poinar, D. Dahl-Jensen, and others. 2007. "Ancient Biomolecules from Deep Ice Cores Reveal a Forested Southern Greenland." *Science* 317(5834):111–114.

Willerslev, E., A. J. Hansen, J. Binladen, T. B. Brand, M. T. P. Gilbert, B. Shapiro, M. Bunce, C. Wiuf, D. A. Gilichinsky, and A. Cooper. 2003. "Diverse Plant and Animal Genetic Records from Holocene and Pleistocene Sediments." *Science* 300:791–795.

Williams, N. 2006. "Footprint Fears for New TB Threat." *Current Biology* 16(19):R821–R822.

Williamson, S. J., D. B. Rusch, S. Yooseph, A. L. Halpern, K. B. Heidelberg, J. I. Glass, C. Andrews-Pfannkoch, D. Fadrosh, C. S. Miller, G. Sutton, and others. 2008. "The Sorcerer II Global Ocean Sampling Expedition: Metagenomic Characterization of Viruses within Aquatic Microbial Samples." *PLoS One* 3(1):e1456.

Willmann, M. R. 2001. "Studying the Historic Migrations of the Irish Potato Famine Pathogen Using Ancient DNA." *Trends in Plant Science* 6(10):450.

Wilson, A. C. and V. M. Sarich. 1969. "A Molecular Time Scale for Human Evolution." *Proceedings of the National Academy of Sciences of the United States of America* 63(4):1088–1093.

Wilson, J. F., D. A. Weiss, M. Richards, M. G. Thomas, N. Bradman, and D. B. Goldstein. 2001. "Genetic Evidence for Different Male and Female Roles during Cultural Transitions in the British Isles." *Proceedings of the National Academy of Sciences of the United States of America* 98(9):5078–5083.

Wirth, T., X. Wang, B. Linz, R. P. Novick, J. K. Lum, M. Blaser, G. Morelli, D. Falush, and M. Achtman. 2004. "Distinguishing Human Ethnic Groups by Means of Sequences from *Helicobacter pylori*: Lessons from Ladakh." *Proceedings of the National Academy of Sciences of the United States of America* 101(14):4746–4751.

Wiwchar, D. 2004. "Nuu-chah-nulth Blood Returns to West Coast." *Ha-Shilth-Sa* 32(25):1–4.

Wolpoff, M. H. 1996. "Interpretations of Multiregional Evolution." *Science* 274(5288):704–707.

Wolpoff, M. H., J. Hawks, and R. Caspari. 2000. "Multiregional, not Multiple Origins." *American Journal of Physical Anthropology* 112(1):129–136.

Wolpoff, M. H., J. N. Spuhler, F. H. Smith, J. Radovčić, G. Pope, D. W. Frayer, R. Eckhardt, and G. Clark. 1988. "Modern Human Origins." *Science* 241(4867):772–773.

Wolpoff, M. H., A. G. Thorne, D. W. Smith, D. W. Frayer, and G. G. Pope. 1994. "Multiregional Evolution: A World-wide Source for Modern Human Populations." In: M. H. Nitecki and D. V. Nitecki, editors. *Origins of Anatomically Modern Humans*. New York: Plenum Press. pp. 175–199.

Wood, E. T., D. A. Stover, C. Ehret, G. Destro-Bisol, G. Spedini, H. McLeod, L. Louie, M. Bamshad, B. I. Strassman, H. Soodyall, and others. 2005. "Contrasting Patterns of Y Chromosome and mtDNA Variation in Africa: Evidence for Sex-based Demographic Processes." *European Journal of Human Genetics* 13(7):867–876.

Wood, J. W., R. J. Ferrell, and S. N. Dewitte-Avina. 2003. "The Temporal Dynamics of the Fourteenth-century Black Death: New Evidence from English Ecclesiastical Records." *Human Biology* 75(4):427–448.

Woodward, S. R., N. Weyand, and M. Bunnell. 1994. "DNA Sequence from Cretaceous Period Bone Fragments." *Science* 266(5188):1229–1232.

Xing, J., W. S. Watkins, A. Shlien, E. Walker, C. D. Huff, D. J. Witherspoon, Y. Zhang, T. S. Simonson, R. B. Weiss, J. D. Schiffman, and others. 2010. "Toward a More Uniform

Sampling of Human Genetic Diversity: A Survey of Worldwide Populations by High-density Genotyping." *Genomics* 96(4):199–210.

Xue, Y., T. Zerjal, W. Bao, S. Zhu, S-K. Lim, Q. Shu, J. Xu, R. Du, S. Fu, P. Li, and others. 2005. "Recent Spread of a Y-chromosomal Lineage in Northern China and Mongolia." *American Journal of Human Genetics* 77(6):1112–1116.

Yanagihara, R., V. R. Nerurkar, I. Scheirich, H. T. Agostini, C. S. Mgone, X. H. Cui, D. V. Jobes, C. L. Cubitt, C. F. Ryschkewitsch, D. B. Hrdy, and others. 2002. "JC Virus Genotypes in the Western Pacific Suggest Asian Mainland Relationships and Virus Association with Early Population Movements." *Human Biology* 74(3):473–488.

Yao, Y-G., Q-P. Kong, H-J. Bandelt, T. Kivisild, and Y-P. Zhang. 2002. "Phylogeographic Differentiation of Mitochondrial DNA in Han Chinese." *American Journal of Human Genetics* 70(3):635–651.

Yen, D. E. 1993. "The Origins of Subsistence Agriculture in Oceania and the Potentials for Future Tropical Food Crops." *Economic Botany* 47(1):3–14.

Yen, D. E. and J. M. Wheeler. 1968. "Introduction of Taro into the Pacific—Indications of the Chromosome Numbers." *Ethnology* 7(3):259–267.

Yindee, M., B. H. Vlamings, W. Wajjwalku, M. Techakumphu, C. Lohachit, S. Sirivaidyapong, C. Thitaram, A. Amarasinghe, P. Alexander, B. Colenbrander, and others. 2010. "Y-chromosomal Variation Confirms Independent Domestications of Swamp and River Buffalo." *Animal Genetics* 41(4):433–435.

Yogo, Y., C. Sugimoto, H-Y. Zheng, H. Ikegaya, T. Takasaka, and T. Kitamura. 2004. "JC Virus Genotyping Offers a New Paradigm in the Study of Human Populations." *Reviews in Medical Virology* 14(3):179–191.

Yule, J. V., C. X. J. Jensen, A. Joseph, and J. Goode. 2009. "The Puzzle of North America's Late Pleistocene Megafaunal Extinction Patterns: Test of New Explanation Yields Unexpected Results." *Ecological Modelling* 220(4):533–544.

Zalloua, P. A., D. E. Platt, M. El Sibai, J. Khalife, N. Makhoul, M. Haber, Y. Xue, H. Izaabel, E. Bosch, S. M. Adams, and others. 2008. "Identifying Genetic Traces of Historical Expansions: Phoenician Footprints in the Mediterranean." *American Journal of Human Genetics* 83(5):633–642.

Zeder, M. A., E. Emshwiller, B. D. Smith, and D. G. Bradley. 2006. "Documenting Domestication: The Intersection of Genetics and Archaeology." *Trends in Genetics* 22(3):139–155.

Zegura, S. L., T. M. Karafet, L. A. Zhivotovsky, and M. F. Hammer. 2004. "High-resolution SNPs and Microsatellite Haplotypes Point to a Single, Recent Entry of Native American Y Chromosomes into the Americas." *Molecular Biology and Evolution* 21(1):164–175.

Zerega, N. J. C., D. Ragone, and T. J. Motley. 2004. "Complex Origins of Breadfruit (*Artocarpus altilis*, Moraceae): Implications for Human Migrations in Oceania." *American Journal of Botany* 91(5):760–766.

Zerjal, T., R. S. Wells, N. Yuldasheva, R. Ruzibakiev, and C. Tyler-Smith. 2002. "A Genetic Landscape Reshaped by Recent Events: Y-chromosomal Insights into Central Asia." *American Journal of Human Genetics* 71(3):466–482.

Zerjal, T., Y. Xue, G. Bertorelle, R. S. Wells, W. Bao, S. Zhu, R. Qamar, Q. Ayub, A. Mohyuddin, S. Fu, and others. 2003. "The Genetic Legacy of the Mongols." *American Journal of Human Genetics* 72(3):717–721.

Zeuner, F. E. 1963. *A History of Domesticated Animals*. London: Hutchinson.

Zhadanov, S. I., M. C. Dulik, M. Markley, G. W. Jennings, J. B. Gaieski, G. Elias, and T. G. Schurr. 2010. "Genetic Heritage and Native Identity of the Seaconke Wampanoag Tribe of Massachusetts." *American Journal of Physical Anthropology* 142(4):579–589.

Zhang, D., G. Rossel, A. Kriegner, and R. Hijmans. 2004. "AFLP Assessment of Diversity in Sweet Potato from Latin America and the Pacific Region: Its Implications on the Dispersal of the Crop." *Genetic Resources and Crop Evolution* 51(2):115–120.

Zhang, L-B., Q. Zhu, Z-Q. Wu, J. Ross-Ibarra, B. S. Gaut, S. Ge, and T. Sang. 2009. "Selection on Grain Shattering Genes and Rates of Rice Domestication." *New Phytologist* 184(3):708–720.

Zhao, Z., N. Yu, Y-X. Fu, and W-H. Li. 2006. "Nucleotide Variation and Haplotype Diversity in a 10-kb Noncoding Region in Three Continental Human Populations." *Genetics* 174(1):399–409.

Zheng, H-X., S. Yan, Z-D. Qin, Y. Wang, J-Z. Tan, H. Li, and L. Jin. 2011. "Major Population Expansion of East Asians Began before Neolithic Time: Evidence from mtDNA Genomes." *PLoS One* 6(10):e25835.

Zhivotovsky, L. A., P. A. Underhill, C. Cinnioğlu, M. Kayser, B. Morar, T. Kivisild, R. Scozzari, F. Cruciani, G. Destro-Bisol, G. Spedini, and others. 2004. "The Effective Mutation Rate at Y Chromosome Short Tandem Repeats, with Application to Human Population-divergence Time." *American Journal of Human Genetics* 74(1):50–61.

Zhivotovsky, L. A. 1999. "Recognition of the Remains of Tsar Nicholas II and His Family: A Case of Premature Identification?" *Annals of Human Biology* 26(6):569–577.

Zink, A. R., C. Sola, U. Reischl, W. Grabner, N. Rastogi, H. Wolf, and A. G. Nerlich. 2003. "Characterization of *Mycobacterium tuberculosis* Complex DNAs from Egyptian Mummies by Spoligotyping." *Journal of Clinical Microbiology* 41(1):359–367.

Zischler, H., M. Höss, O. Handt, A. von Haeseler, A. van der Kuyl, and J. Goudsmit. 1995. "Detecting Dinosaur DNA." *Science* 268(5214):1192–1193.

Zizumbo-Villarreal, D. and P. Colunga-GarcíaMarín. 2010. "Origin of Agriculture and Plant Domestication in West Mesoamerica." *Genetic Resources and Crop Evolution* 57(6):813–825.

Zuckerkandl, E. 1963. "Perspectives in Molecular Anthropology." In: S. Washburn, editor. *Classification and Evolution*. Chicago: Aldine. pp. 243–272.

Zuckerkandl, E. and L. Pauling. 1965. "Molecules as Documents of Evolutionary History." *Journal of Theoretical Biology* 8(2):357–366.

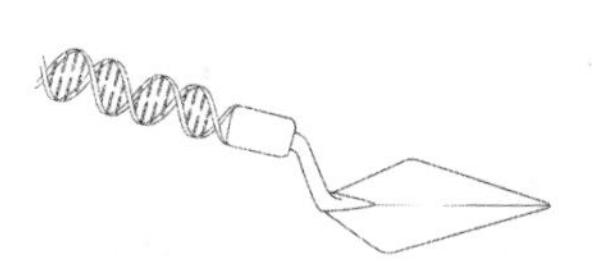

Index

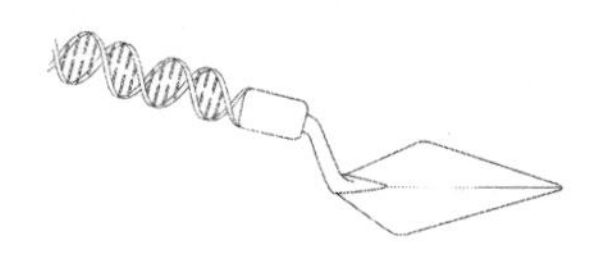

About the Authors

Elizabeth (Lisa) Matisoo-Smith is Professor of Biological Anthropology at the University of Otago and a Principal Investigator in the Allan Wilson Centre for Molecular Ecology and Evolution and with National Geographic's Genographic Project. Her main research focus is on using DNA data to address issues in Pacific prehistory. She developed the use of the commensal model, looking at ancient and modern DNA variation in transported animals, to track human migration pathways in the Pacific. She now combines this approach with analyses of ancient and modern human DNA, working closely with local communities and national museums in the Pacific region. Lisa has undertaken fieldwork in California, France, French Polynesia, New Guinea, Tokelau, New Zealand, and most recently, Chile.

K. Ann Horsburgh is a Research Fellow in the Department of Anatomy at the University of Otago and an Honorary Research Fellow in the School of Geography, Archaeology, and Environmental Studies at the University of the Witwatersrand. She earned her PhD in Anthropology at Stanford University where she wrote her dissertation on cattle, dog, and sheep domestication in southern African prehistory using ancient DNA. She is currently developing the application of next generation sequencing technology to better understand the processes of human selection in cattle domestication across Africa.

Left Coast Press, Inc. is committed to preserving ancient forests and natural resources. We elected to print this title on 30% post consumer recycled paper, processed chlorine free. As a result, for this printing, we have saved:

1 Trees (40' tall and 6-8" diameter)
1 Million BTUs of Total Energy
167 Pounds of Greenhouse Gases
908 Gallons of Wastewater
61 Pounds of Solid Waste

Left Coast Press, Inc. made this paper choice because our printer, Thomson-Shore, Inc., is a member of Green Press Initiative, a nonprofit program dedicated to supporting authors, publishers, and suppliers in their efforts to reduce their use of fiber obtained from endangered forests.

For more information, visit www.greenpressinitiative.org

Environmental impact estimates were made using the Environmental Defense Paper Calculator. For more information visit: www.papercalculator.org.